Rethinking Engineering Education

The CDIO Approach

Rethinking Engineering Education

The CDIO Approach

Edward F. Crawley
Massachusetts Institute of Technology

Johan Malmqvist
Chalmers University of Technology

Sören Östlund
KTH - Royal Institute of Technology

Doris R. Brodeur
Massachusetts Institute of Technology

Edward F. Crawley
Massachusetts Institute of Technology
77 Massachusetts Avenue – 33-409
Cambridge, MA 02139
USA

Johan Malmqvist
Department of Product and Production Development
Chalmers University of Technology
SE – 412 96 Göteborg
SWEDEN

Sören Östlund
Department of Solid Mechanics
KTH – Royal Institute of Technology
SE – 100 44 Stockholm
SWEDEN

Doris R. Brodeur
Massachusetts Institute of Technology
77 Massachusetts Avenue – 37-391
Cambridge, MA 02139
USA

Library of Congress Control Number: 2007921087

ISBN 978-0-387-38287-6 e-ISBN 978-0-387-38290-6

Printed on acid-free paper.

9 8 7 6 5 4 3 2 1

springer.com

Table of Contents

Educating Engineers for 2020 and Beyond

Charles M. Vest
President Emeritus
MIT

Most of my career was played out in the 20th century – the century of physics, electronics, and high speed communications and transportation. And now, we all – and especially our students –have the privilege of living through the transition to the 21st century – presumably the century of biology and information.

As this transition occurs, it is an appropriate time to rethink engineering education. When I look back over my 35-plus years as an engineering educator, I realize that many things have changed remarkably, but others seem not to have changed at all. Challenges that have been with us for the past 35 years include making the first university year more exciting, communicating what engineers actually do, and bringing the richness of human diversity into the engineering workforce. Students must learn how to merge the physical, life, and information sciences at the nano-, meso-, micro- and macro- scales; embrace professional ethics and social responsibility, be creative and innovative, and write and communicate well. Our students should be prepared to live and work as global citizens, understand how engineers contribute to society. They must develop a basic understanding of business processes; be adept at product development and high-quality manufacturing; and know how to *conceive, design, implement and operate* complex engineering systems of appropriate complexity. They must increasingly do this within a framework of sustainable development, and be prepared to live and work as global citizens. That is a tall order ... perhaps even an impossible order.

But is it really? I meet students in the hallways of MIT and other universities who can do all of these things—and more. So, we must keep our sights high. But how are we going to accomplish all this teaching and learning? What has stayed constant, and what needs to be changed?

As we think about the challenges ahead, it is important to remember that some things are constant. Students, for example, are driven by passion, curiosity, engagement, and dreams. Although we cannot know exactly what they should be taught, we can focus on the environment and context in which they learn, and the forces, ideas, inspirations, and empowering authentic situations to which they are exposed.

Another constant is the need for students to acquire a sound basis in science, engineering principles, and analytical capabilities. In my view, a deep understanding of the fundamentals is still the most important thing we provide. Much of our current view of the engineering fundamentals was shaped by what is commonly termed the "engineering science revolution." This revolution was spawned largely by faculty at MIT who, building on their experiences gained by developing radar systems during World War II, created a radically different way to practice and teach engineering. A towering legacy of this era, with contributions from many major universities, was a new world of engineering education that was built on a solid foundation of science more than on traditional macroscopic phenomenology, charts, handbooks, and codes. The new engineering science required a new panoply of textbooks and laboratories. However,the creators of this new vision of engineering education did not mean to displace the excitement of engineering, the opportunity for students to design and build, or the need for teamwork and ethics, meant to enrich the student experience. Along the way, something got lost. We need to rethink engineering education, and find a new balance.

Perhaps I am so old fashioned I still believe that masterfully conceived, well-delivered lectures are still wonderful teaching and learning experiences. They still have their place. But even I admit there is a good deal of truth in what my extraordinary friend, Murray Gell-Mann, Winner of Nobel Prize in Physics, 1929 likes to say, "We need to move from the sage on the stage to the guide on the side." Studio teaching, team projects, open-ended problem solving, experiential learning, engagement in research, should be integral elements of engineering education.

The philosophy of the CDIO approach to engineering education captures these essential features of a modern engineering education - excitement about what engineers do, deep learning of the fundamentals, skills, and the knowledge of how engineers contribute to society. It is taught in a way that captures our students' passion.

I encourage you to read about this integrated approach, and consider how it might influence the practice of engineering education at your university.

CHAPTER ONE INTRODUCTION

RATIONALE

The purpose of engineering education is to provide the learning required by students to become successful engineers—technical expertise, social awareness, and a bias toward innovation. This combined set of knowledge, skills, and attitudes is essential to strengthening productivity, entrepreneurship, and excellence in an environment that is increasingly based on technologically complex and sustainable products, processes, and systems. It is imperative that we improve the quality and nature of undergraduate engineering education.

In the last two decades, leaders in academia, industry, and government began to address the necessity for reform by developing views of the desired attributes of engineers. Through this endeavor, we identified an underlying critical need—to educate students who are able to Conceive-Design-Implement-Operate complex, value-added engineering products, processes and systems in a modern, team-based environment. It is from this emphasis on the product, process, or system lifecycle that the initiative derives its name–CDIO.

Within these pages, we demonstrate how conceiving, designing, implementing, and operating products, processes, and systems is the appropriate context for engineering education. The CDIO approach builds on stakeholder input to identify the learning needs of the students in a program, and construct a sequence of integrated learning experiences to meet those needs. We incorporate a comprehensive and broadly applicable approach to improving curriculum, teaching and learning, and workspaces that is supported by robust assessment and change processes. By these means, we seek to significantly improve the quality and nature of undergraduate engineering education.

BACKGROUND

In the 1980s and 1990s, engineers in industry and government, along with university program leaders, began to discuss improvements in the state of engineering education. In this process, they considered the proficiencies of engineering graduates of recent years and developed lists of the desired attributes of engineers. Common among these lists was an implicit criticism of current engineering education for prioritizing the teaching of theory, including mathematics, science, and technical disciplines, while not placing enough emphasis on laying the foundation for practice, which emphasizes skills such as design, teamwork, and communications.

This criticism reveals the tension between two key objectives within contemporary engineering education: the need to educate students as *specialists* in a range of technologies—each with increasing levels of knowledge required for professional mastery—while at the same time teaching students to develop as *generalists* in a range of personal, interpersonal, and product, process, and system building skills.

Engineering programs in many parts of the world that exemplify this tension are the products of the evolution of engineering education in the last half century. Through those years, programs moved from a practice-based curriculum to an engineering science-based model. The intended consequence of this change was to offer students a rigorous, scientific foundation that would equip them to address unknown future technical challenges. The unintended consequence of this change was a shift in the culture of engineering education that diminished the perceived value of key skills and attitudes that had been the hallmark of engineering education until that time. Thus evolved the tension between theory and practice.

The challenge that remains is that of introducing change to relieve this tension, to respond to the needs of our external stakeholders, to reform our programs and educational approaches, and in fact, to transform the culture of education.

THE CDIO INITIATIVE

The CDIO Initiative meets this challenge by educating students as well-rounded engineers who understand how to Conceive-Design-Implement-Operate complex, value-added engineering products, processes, and systems in a moderny, team-based environment. The Initiative has three overall goals: *To educate students who are able to:*

- *Master a deeper working knowledge of technical fundamentals.*
- *Lead in the creation and operation of new products, processes, and systems.*
- *Understand the importance and strategic impact of research and technological development on society.*

This education stresses the fundamentals, and is set in the *context* of conceiving, designing, implementing, and operating products, processes, and systems. We seek to develop programs that are educationally effective *and* more exciting to students, attracting them to engineering, retaining them in the program and in the profession.

This context of conceiving, designing, implementing, and operating is appropriate both because *it is* the professional role of engineers and because it provides the natural setting in which to teach key pre-professional engineering skills and attitudes. Within that context, we develop an integrated approach to identifying students' learning needs and construct a sequence of learning experiences to meet them.

The essential feature of the CDIO approach is that it creates *dual-impact* learning experiences that promote deep learning of technical fundamentals and of practical skill sets. We use modern pedagogical approaches, innovative teaching methods, and new learning environments to provide real-world learning experiences. These concrete learning experiences create a cognitive framework for learning the abstractions associated with the technical fundamentals, and provide opportunities for active application that facilitates understanding and retention. Thus they provide the pathway to deeper working knowledge of the fundamentals. These concrete experiences also impart learning in personal and interpersonal skills, and product, process, and system building skills.

THE SYLLABUS AND THE STANDARDS

A rigorous engineering process has been applied to the design of the CDIO approach to ensure that it achieves its goals. We build an integrated approach to identifying the learning needs of the students in a program, and to construct a sequence of learning experiences to meet those needs. These two elements are captured in a best-practice framework, consisting of the CDIO Syllabus and the CDIO Standards.

Specific learning outcomes are codified in the CDIO Syllabus. The Syllabus is a rational, relevant, and consistent set of skills for an engineer. The Syllabus was derived from needs assessment and source documents, and tested by peer review. The proficiency expectations for graduating students are set with stakeholder input. These learning outcomes then form the basis for program design and assessment.

A CDIO program creates a curriculum organized around mutually supporting technical disciplines with personal and interpersonal skills, and product, process, and system building skills highly interwoven. These programs are rich with student design-implement experiences conducted in modern workspaces. They feature active and experiential learning and are continuously improved through a robust, quality assessment process. These characteristics are formalized in twelve CDIO Standards, which define the distinguishing features of a CDIO program; serve as guidelines for educational program reform and evaluation; create benchmarks and goals with worldwide application; and provide a framework for continuous improvement.

IMPLEMENTATION AND EVOLUTION

Development and implementation of the CDIO approach was initiated at four universities: Chalmers University of Technology (Chalmers) in Göteborg, the Royal Institute of Technology (KTH) in Stockholm, Linköping University (LiU) in Linköping, and the Massachusetts Institute of Technology (MIT) in Cambridge, Massachusetts. The number of programs collaborating in the Initiative has expanded to more than 20 universities worldwide.

Little in our approach has been invented of whole cloth. We have built upon research and best practices found within our collaborating universities and many other universities around the world who are seeking to improve engineering education. Many have made important contributions. The CDIO Initiative seeks to build on and systematize this international body of work, to develop a set of broadly applicable shared approaches and open-source resources that guide and accelerate engineering education reform. We recognize that for most programs, extensive financial and personal resources are not available. We use the shared open-source resources and parallel-coordinated efforts to facilitate rapid transition to a steady state that largely retasks existing personnel, time, and space resources.

Nothing in our approach is prescriptive. The CDIO approach must be adapted to each program—its goals, university, national, and disciplinary contexts. It is aligned with many other movements for educational change, but unlike national accreditation and assessment standards that state objectives, we provide a pallet of potential solutions to the comprehensive reform of engineering education. Many programs around the world are working on aspects of this issue and making important contributions. Many have already developed along the lines of the twelve CDIO Standards independently. We recognize this. We invite you to share your results, and contribute to our collective effort.

THE BOOK

We have written this book to serve as an introduction to the approaches and resources created by the CDIO Initiative. It is a practical guide with enough information to acquaint you with the high-level rationale, philosophy, and key approaches, and how they have evolved in a historical and societal context. The book points to more detailed resources that are contained in other publications, in workshops, and on the web.

Chapter 2 continues with an in-depth overview of the CDIO Initiative. This chapter will leave the reader with an understanding of the need for change, the goals, vision, and pedagogical foundation of the CDIO approach, and the essential elements of implementation. Chapter 3 explains the process for identifying the desired skills of an engineer and the learning outcomes for students in a program. Chapters 4 through 6 then describe in some detail the curricular, workspace, and teaching and learning aspects of the approach. Chapters 7 through 9 discuss program evaluation, student assessment, and implementation and change processes. The book concludes with a historical perspective of engineering education, in order to provide the reader with the background to understand the context of change, and an informed outlook to the future.

CHAPTER TWO
OVERVIEW

INTRODUCTION

The objective of engineering education is to educate students who are "ready to engineer," that is, broadly prepared with the pre-professional skills of engineering, and deeply knowledgeable of the technical fundamentals. It is the task of engineering educators to continuously improve the quality and nature of undergraduate engineering education in order to meet this objective. Over the past 25 years, many in industry, government, and university programs have addressed the need for reform of engineering education, often by stating the desired outcomes in terms of attributes of engineering graduates. By examining these views, we identified an underlying need: to educate students to understand how to Conceive-Design-Implement-Operate complex value-added engineering products, processes and systems in a modern, team-based environment.

The CDIO approach reforms engineering education to meet this underlying need. The value of this approach to students is built on three premises, which reflect its goals, vision, and pedagogical foundation:

- That the underlying need is best met by setting goals that stress the fundamentals, while at the same time making the process of conceiving-designing-implementing-operating products, processes, and systems the context of engineering education.
- That the learning outcomes for students should be set through stakeholder involvement, and met by constructing a sequence of integrated learning experiences, some of which are experiential, that is, they expose students to the situations that engineers encounter in their profession.
- That proper construction of these integrated learning activities will cause the activities to be *dual-impact*, facilitating student learning of critical personal and interpersonal skills, and product, process, and system building skills, and simultaneously enhancing the learning of the fundamentals.

The CDIO approach incorporates comprehensive and broadly applicable processes for improving curriculum, teaching and learning, and workspaces, and is supported by robust assessment, and change processes.

This overview chapter outlines the key premises and features of the CDIO Initiative. It begins with a discussion of the motivation for improvement in engineering education, including a discussion of the needs of our students, the historical environment of our education, and the requirements for an effective program of reform. The second section describes the Initiative in some detail: its goals, vision, and pedagogical foundation. The structure of this second section serves as the framework for many of the remaining chapters of the book, which go into more detail on the topics of setting goals for learning, improving curriculum and workspaces, teaching and learning, and conducting student assessment and program evaluation. The final part of the chapter describes approaches to development, including the available resources and collaboration approach, and underscores the need to recognize educational reform as a process for organizational and cultural change at the university.

CHAPTER OBJECTIVES

This chapter is designed so that you can

- recognize the contemporary motivation for engineering education reform
- explain the underlying goals, vision, and pedagogical foundation
- describe the key characteristics of a CDIO program
- explain the approach to the development of the CDIO Initiative

MOTIVATION FOR CHANGE

Engineers build things that serve society. To quote Theodore von Kármán [1], "*Scientists discover the world that exists; engineers create the world that never was.*" The 1828 charter of the Institution of Civil Engineers [2] states that engineering is "the art of directing great sources of power in nature for the use and convenience of man." Creation of new products and direction of natural resources remain the tasks of engineers today.

What modern engineers do

Modern engineers are engaged in all phases of the lifecycle of products, processes and systems that range from the simple to the incredibly complex, but all have one feature in common. They meet a need of a member of society. Good engineers observe and listen carefully to determine the needs of the member of society for whom the benefit is intended. They are involved in conceiving the device or system.

Modern engineers design products, processes, and systems that incorporate technology. Sometimes this is state-of-the-art technology, pushing new frontiers, and creating new capabilities. That is the stuff of startups and breakthrough innovations. However, much of engineering design is performed by applying and adapting existing technology to meet society's changing needs. In most of the world, society is uplifted by broad-based applications of existing technology. Good engineers apply appropriate technology to design.

Engineers lead, and in some cases, execute the implementation of the design to actual realization of the product, process, or system. All engineers should design so that their systems are implemented easily and in a sustainable way. Some engineers, such as those who develop software, are actually involved in both the design and implementation of the code. In other industries, engineers specialize in implementation, such as manufacturing engineers.

Modern engineers work in teams when they conceive, design and implement the product, process, or system. Teams are often geographically distributed and international. Engineers exchange thoughts, ideas, data and drawings, elements and devices with others around the work site and around the world. They capture the tacit knowledge of a system's design and implementation so that it can be revised and upgraded in the future. Good engineers work in teams and communicate effectively, while always exercising personal creativity and responsibility.

In order to deliver a benefit to a member of society, engineering devices and systems must be operated. Simpler devices, such as, stoves, cars, or laptop computers, are operated by private users. More complex systems, such as, industrial furnaces, aircraft, or communication networks, are operated by professionals. Good engineers consider and plan for the operation of the product, process, or system as an integral part of design. They are sometimes involved in the operation of the system as well.

Conceive-Design-Implement-Operate

Modern engineers lead or are involved in all phases of a product, process, or system lifecycle. That is, they Conceive, Design, Implement, and Operate. The *Conceive* stage includes defining customer needs; considering technology, enterprise strategy, and regulations; and developing conceptual, technical, and business plans. The second stage, *Design*, focuses on creating the design, that is, the plans, drawings, and algorithms that describe what product, process, or system will be implemented. The *Implement* stage refers to the transformation of the design into the product, including hardware manufacturing, software coding, testing, and validation. The final stage, *Operate*, uses the implemented product, process, or system to deliver the intended value, including maintaining, evolving, recycling, and retiring the system.

These four terms, and the activities and outcomes of the four phases, have been chosen because they are applicable to a wide range of engineering disciplines. Details of the tasks that fall into these four main phases—conceiving, designing, implementing, and operating—are found in Figure 2.1. Note that sequence is not strictly implied by the figure. For example, in spiral development models of product development, there is a great deal of iteration among

Conceive		Design		Implement		Operate	
Mission	**Conceptual Design**	**Preliminary Design**	**Detailed Design**	**Element Creation**	**Systems' Integration & Test**	**Lifecycle Support**	**Evolution**
• Business Strategy • Technology Strategy • Customer Needs • Goals • Competitors • Program Plan • Business Plan	• Requirements • Function • Concepts • Technology • Architecture • Platform Plan • Market Positioning • Regulation • Supplier Plan • Commitment	• Requirements Allocation • Model Development • System Analysis • System Decomposition • Interface Specifications	• Element Design • Requirements Verification • Failure & Contingency Analysis • Validated Design	• Hardware Manufacturing • Software Coding • Sourcing • Element Testing • Element Refinement	• System Integration • System Test • Refinement • Certification • Implementation Ramp-up • Delivery	• Sales & Distribution • Operations • Logistics • Customer Support • Maintenance & Repair • Recycling • Upgrading	• System Improvement • Product Family Expansion • Retirement

FIGURE 2.1. CONCEIVING - DESIGNING - IMPLEMENTING - OPERATING AS A LIFECYCLE MODEL OF A PRODUCT, PROCESS, PROJECT, OR SYSTEM

these tasks. Yet, whatever the sequence, these tasks are completed in most successful product developments, and therefore, form the core processes executed by engineers in building products, processes, and systems that meet the needs of society.

The most obvious mapping of these four phases is onto the development of discrete electro/mechanical/information products and systems in serial production, such as cars, aircraft, ships, software, computers, and communications devices. Manufacturing engineers actually plan, design, realize, and operate the manufacturing processes for these discrete products and systems. Other engineers envision, design, develop, and deploy networks and systems of these devices, including transportation networks and communication systems. In software, engineers envision, design, write, and operate code. In chemical engineering and similar process industries, engineers conceive, design, build, and operate a plant or facility. In civil engineering, similar steps are taken for the planning, design, construction, and operation of a single project.

Appropriately interpreted, this common paradigm of conceiving, designing, implementing, and operating covers the essential professional activities of the vast majority of engineers. In order to simplify and standardize the terminology in this book, the terms *product, process, and system* are consistently used for the object the engineer designs and implements, which, depending on the sector, is called a product, process, system, device, network, code, plant, facility, or project. Likewise *conceive, design, implement, and operate* are consistently used for the four major tasks in realizing these products, processes, and systems. As a shorthand, this lifecycle process is sometimes simply called *system building*.

The need for reform of engineering education

The task of higher education is to educate students to become effective modern engineers—able to participate and eventually to lead in aspects of conceiving, designing, implementing, and operating systems, products, processes, and projects. To do this, students must be technically expert, socially responsible, and inclined to innovate. Such an education is essential for achieving productivity, entrepreneurship, and excellence in an environment that is increasingly based on technologically complex systems that must be sustainable. It is widely acknowledged that we must do a better job at preparing engineering students for this future, and that we must do this by systematically reforming engineering education. The better preparation of engineering students through systematic reform of engineering education is the ultimate intent of the CDIO Initiative.

Any approach to improving engineering education must address two central questions:

- *What is the full set of knowledge, skills, and attitudes that engineering students should possess as they leave the university, and at what level of proficiency ?*
- *How can we do better at ensuring that students learn these skills?*

These are essentially the *what* and *how* questions that engineering educators commonly face. Focusing on the first question, there is a seemingly irreconcilable tension between two positions in engineering education. On one hand, there is the need to convey the ever-increasing body of technical knowledge that graduating students must master. On the other hand, there is growing acknowledgment that engineers must possess a wide array of personal and interpersonal skills; as well as the product, process, and system building knowledge and skills required to function on real engineering teams to produce real products and systems.

This tension is manifest in the apparent difference of opinion between engineering educators and the broader engineering community that ultimately employs engineering graduates. University-based engineers traditionally strike a balance that emphasizes the importance of a body of technical knowledge. However, beginning in the late 1970s and early 1980s, and increasingly in the 1990s, industrial representatives began expressing concern about this balance, articulating the need for a broader view that gives greater emphasis to the personal and interpersonal skills; and product, process, and system building skills. The Finiston Report of 1978 in the United Kingdom is an early example of this reaction [3]. A few years later in 1984, Bernard M. Gordon, the inventor of the analog-to-digital converter, winner of the U.S. National Medal of Technology, and benefactor of the Gordon Prize for Engineering Education of the U.S. National Academy of Engineering, stated bluntly that "society . . . around the world . . . is not entirely pleased with the current state of general [engineering] education" [4]. Box 2.1 is an excerpt of his address to the annual conference of the European Society for Engineering Education (SEFI).

By the 1990's, this trend of criticizing university engineering education spread widely. For example, The Boeing Company in the United States organized an effort to influence university engineering education by setting forth its list of desired attributes of an engineer [5], as listed in Box 2.2. More broadly, the reac-

Box 2.1. **What is an Engineer?**

It is apparent that society around the world, particularly, the western world, is not entirely pleased with the current state of general education. Its displeasure is reflected in the barrage of criticism leveled at the graduate who cannot read effectively, cannot write effectively, and cannot master moderately complex arithmetic. The well-publicized question, "Why can't Johnny read?" sums up the societal concerns.

A parallel question, "Why can't Mr. /Dr. Engineer engineer effectively?" is now increasingly being asked, and sums up the frustration of engineering supervisors and of the public who suffer from the failures of inadequate designs. Critics of engineering education often cite the following inadequacies among the complaints about the educational system's "product":

- Disproportionately low and increasingly poor economic return for the amount of employed engineering resources
- Limited formal training in, and exposure to, a breadth of basic technical knowledge
- Inadequate training and orientation to a meaningful depth of engineering skills
- Inadequate understanding of the importance of precise test and measurement
- Insufficient competitive drive and perseverance
- Inadequate communication skills
- Lack of discipline and control in work habits
- Fear of taking personal risks

Therefore, it is appropriate that we re-examine our perceptions of real engineering to focus our attention on the content in terms of what we want engineers to do in their careers, while we are exploring the application of new technology to the methods of education.

Definition

I propose to define a REAL, that is, professional, ENGINEER as *one who has attained and continuously enhances technical, communications, and human relations knowledge, skills, and attitudes, and who contributes effectively to society by theorizing, conceiving, developing, and producing reliable structures and machines of practical and economic value.*

The greater the breadth of knowledge, the more varied and accomplished the skills, and the more dedicated the attitude of any individual engineer, the more significant will be the accomplishment, resulting in proper recognition as a role model, teacher, and leader. . . .

Knowledge

Knowledge for a real engineer is more than acquired data, and certainly much more than acquired engineering data. The cognitive process is different from the acquisitive process. While today's engineer may use information technology to make any of the world's data instantly available, the real engineer has developed a relational understanding of the data and will have learned how to recall and correlatively process relevant data in order to synthesize new information to solve problems.

The areas of required knowledge are not limited to those of science or technology, as consideration of the role of the engineer as leader will reveal. An understanding of societal evolution through study of history, economics, sociology, psychology, literature, and arts will enhance the value of the engineering contribution. And, in the shrinking world that the new communications technology is producing, we should not forget the study of foreign languages—an item often ignored on the western side of the Atlantic.

Skills

A real engineer's skills are essentially scheduled problem solving techniques of design, in which the concentrated disciplines of science and technology are exercised with the personal creativity and judgment developed from training and experience. In addition, because engineering accomplishments are achieved in a group environment, communication skills are critical to the roles as follower and as leader.

These skills can be acquired only by doing: the practice may be on simulated problems, or, as for the entry-level medical doctor, on real cases under expert supervision. However, no

(*Continued*)

Box 2.1. What is an Engineer?—Cont'd

amount of case study can replace the practice in learning how to debug a design, for example. The case study technique may be useful, but it is not sufficient to qualify the real engineer.

Attitudes

A real engineer's attitudes will directly affect the quality of his design solutions, whatever the problem. The real engineer is a leader of a team of resources: financial, personal, and material, at all levels of engineering activity. Successful team leadership implies a degree of self-criticism, where egotism and humility have counterbalancing influences. It requires a spirit of curiosity and courage that leads to creativity and innovation. Successful leadership is characterized by a forcefulness that gives orders, as well as receives orders, and accepts the challenges of competition in the marketplace with a perseverance to succeed. Leadership exhibits a loyalty downward as well as loyalty upward, and requires the earning of respect of project team members for personal competence, tolerance, and supervisory guidance.

– B. M. Gordon, Analogic Corporation

Box 2.2. Desired Attributes of an Engineer

- A good understanding of engineering science fundamentals
  - Mathematics (including statistics)
  - Physical and life sciences
  - Information technology (far more than computer literacy)
- A good understanding of design and manufacturing processes
- A multi-disciplinary, systems perspective
- A basic understanding of the context in which engineering is practiced
  - Economics (including business practices)
  - History
  - The environment
  - Customer and societal needs
- Good communication skills
  - Written, oral, graphic, and listening
- High ethical standards
- An ability to think both critically and creatively—independently and cooperatively
- Flexibility, *i.e.*, the ability and self-confidence to adapt to rapid or major change
- Curiosity and a desire to learn for life
- A profound understanding of the importance of teamwork.

– The Boeing Company

*Reprinted with kind permission of Boeing Management Company.

tion of industry in the developed world included industry-led workshops and programs on engineering education, and industry influence on accrediting and professional bodies. It also included direct industry and foundation funding of educational initiatives, and industry influence on government to create resources and incentives for change. This was not a random or ill-coordinated effort, but a coherent reaction to what industry considered a major threat to its human resource flow from universities. What these and other commentaries by industrialists have in common is that they always underscore the importance of engineering science fundamentals and engineering knowledge, but then go on to list a wider array of skills that typically include elements of design, communications, teamwork, ethics, and other personal skills, and attributes.

Requirements for the reform of engineering education

In response to this input from our stakeholders, we began developing the CDIO Initiative by examining these sources of advice from industry that reflected on the needs for the education of our students. When we tried to synthesize these "lists" that were proposed by industry, we observed that they were driven by a more basic need, that is, the reason society needs engineers in the first place.

Therefore, the starting point of our effort was a restatement of the underlying *need* for engineering education. We believe that every graduating engineer should be able to:

> *Conceive-Design-Implement-Operate complex value-added engineering products, processes, and systems in a modern, team-based environment*

More simply, we must educate engineers who can engineer. For the responsibilities of engineering are these: to execute a sequence of *tasks*, in order to design and implement a *product, process, or system* within an *organization*. This emphasis on the product or system lifecycle (Conceive-Design-Implement-Operate) gives the initiative its name. We define *value-added* as the additional worth created at a particular stage of production, or through image and marketing. It refers to the contribution of the factors of production to raising the value of a product, process, or system.

Conceiving-Designing-Implementing-Operating as the context of engineering education. We assert that conceiving-designing-implementing-operating should be the *context* of engineering education. The *context* for education is the cultural framework, or environment, in which technical knowledge and skills are learned. The culture of the education, the skills we teach, and the attitudes we convey should all indicate that conceiving-designing-implementing-operating is the role of engineers in their service to society. It is important to note that we assert that the product or system lifecycle should be the *context*, not the *content*, of the engineering education. Not every engineer should specialize in product development. Rather, engineers should be educated in disciplines, that is, mechanical, electrical, chemical, or even engineering science. However, they should be educated in those disciplines in a context that will give them the skills and attitudes to be able to design and implement things. This leads us to the first requirement for a program in engineering education reform:

> *The program adopts the principle that product, process, and system development and deployment—conceiving, designing, implementing and operating—are the context for engineering education.*

Later in this chapter we identify this requirement as CDIO Standard 1.

If we accept this conceive-design-implement-operate premise as the *context* of engineering education, we can then rationally derive more detailed learning outcomes for the education of our students. We can systematically answer the first of the two central questions, namely, *"What is the full set of knowledge, skills, and attitudes that engineering students should possess as they leave the university, and at what level of proficiency should they possess them?"*

The rationale for adopting the principle that the system lifecycle—conceiving, designing, implementing and operating—is the appropriate context for engineering education is supported by the following arguments:

- It is what engineers do.
- It is the underlying need and basis for the "skills lists" that industry proposes to university educators.
- It is the natural context in which to teach these skills to engineering students.

The first point has been argued above—what modern engineers do is engage in some or all phases of conceiving, designing, implementing, and operating. The second point is evidenced by the widespread, consistent and organized reaction from industry in the last few decades. The third point is more subtle. In principle, it is possible to teach students the skills and attitudes of engineering while they work by themselves on engineering theory, but this may not be very effective. What could be a more natural way to educate students in these skills than to set the education in the context of product and system development and deployment, that is, the very context in which students will use the skills?

This observation seems so self-evident that it bears consideration as to why the engineering product, process, and system lifecycle is not currently the common context of engineering education. Quite simply, it is that engineering schools are not, by and large, populated by engineer practitioners, but by engineering researchers. These researchers develop engineering science knowledge by conducting research with a reductionist approach that largely rewards the efforts of individuals. In contrast, in the desired near real-life engineering context, the focus is on producing engineering products and systems by conducting development with an integrative approach that largely rewards team efforts. At the same time, this desired context must still emphasize a rigorous treatment of the engineering fundamentals. Consequently, what we must recognize is that the transformation of the education from the current to the desired context is one of cultural change. We must improve both the skills and attitudes of current engineering faculty by enhancing their collective faculty competence.

Some would argue that such a transformation is unimaginable in a university setting. In fact, the current tension in engineering education in many countries is the result of just such a transformation. As recently as the 1950s, and more recently in some countries, university engineering faculty were distinguished practitioners of engineering. Education was based largely on practices and preparation for practice. The 1950s saw the beginning of the engineering science revolution, and the hiring of a cadre of young engineering scientists. The 1960s might be called the golden era, in which students were educated by a mix of the older practice-based faculty and the younger engineering scientists. However, by the 1970s, as older practitioners retired, they were replaced by engineering scientists. On average, the culture and context of engineering education took a pronounced swing toward engineering science.

Maintaining the fundamentals while strengthening the skills. The intended consequence of this change in context and culture that occurred in the latter half of the twentieth century was to place the education of engineering students on a more rigorous and scientific foundation, equipping them to address unknown future technical challenges. Nothing proposed here is intended to minimize the importance of this change, or the vast contributions that engineering science research has produced in the last half-century. However, the unintended consequence of this change was a shift in the culture of engineering education that diminished the perceived value of many of the key skills and attitudes that had been the hallmark of engineering education up to that time. It is not a coincidence, therefore, that in much of the developed world, the late 1970s and 1980s became the period in which industry started to recognize the change in the knowledge, skills, and attitudes of graduating students. Industry reacted in the 1980s with observations and expressions of concern, and when these did not bring results, with a more cohesive response in the 1990s, as previously discussed.

This evolution of engineering faculty composition can also be traced to a notional representation of the way in which a balance was struck between the teaching of personal, interpersonal, and process skills, and product and system building skills; and the technical fundamentals. Figure 2.2 illustrates this evolution. Prior to 1950, the context of practice prevailed. By the 1960s, more balance was prevalent. By the 1980s, engineering science dominated with a strong emphasis on technical fundamentals. The trend is shown as a trade-off curve because, assuming that education is an information transferring activity,

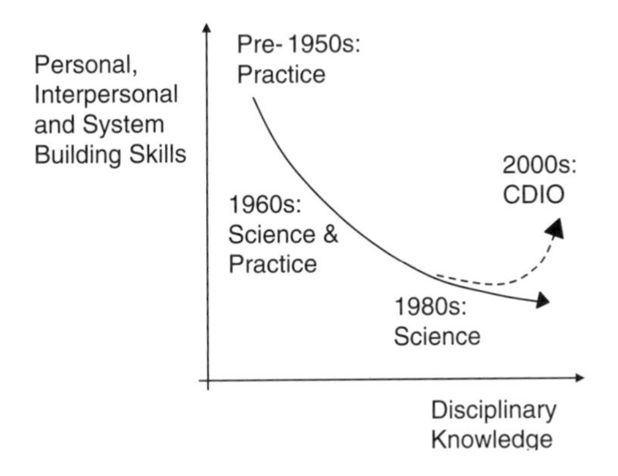

FIGURE 2.2. EVOLUTION OF ENGINEERING EDUCATION

limitations on bandwidth and time allow only a certain amount of content to be covered. This model forces questions such as "What must be removed to make room for this new material?" We assert that there are alternative educational models to that of information transfer that allow relief from this apparent conflict. We can therefore identify the second requirement for successful engineering education reform:

The education emphasizes the technical fundamentals, while strengthening the learning of personal and interpersonal skills; and product, process, and system building skills.

Engagement of key stakeholders. Engineering education has four key stakeholder groups: students, industry, university faculty, and society. To this point, we have considered industry as a major stakeholder of education. Industry is the ultimate customer for the students we graduate, but the immediate customers for the education are the students themselves. Students pass the conventional economic tests for the true customer, that is, they pay for the service of education (or, in some countries, society pays for them), and they are the entity to which the service of education is transferred. In their educational choices, students act as both consumers and investors. They exhibit investor behavior in that they think about the long-term personal and economic impact of a specific course of study. They exhibit consumer behavior if when faced with two options that have equal long-term benefit, they will opt for the more interesting, lower-effort, or enjoyable option. Students are the direct customers and beneficiaries of the educational service and the arbiters of consumer needs, but are often not sufficiently mature or informed in their opinions about the investor aspect of education. Industry, including program alumni working in industry, is informed about investments required for long-term benefit and is therefore a proxy for the investor interests of the students.

University faculty are the developers and deliverers of the knowledge, skills, and attitudes, and they bring their own insights into both the investor and consumer needs of students. In addition to industry, society, through legislation and accreditation, sets requirements on engineering education, including degree requirements and emphasis on societal goals such as sustainable development. In some countries, the government pays students' educational fees. Thus, all four stakeholder groups have important views on educational goals. These factors lead to the third requirement for successful engineering education reform:

The learning outcomes of students in a program should be set in a way that reflects the viewpoints of all key stakeholder groups: students, industry, university faculty, and society.

Attracting and retaining qualified students. Why should industry and engineering educators care about the consumer and investor behaviors of students? In many developed and developing nations, there is a shortage of students in engineering, science, and technology. Students are not attracted to study these fields at university; they are not retained in the programs; or, upon graduation, they move to different fields. All other factors being equal, it would be desirable to reform engineering education so that it was more attractive to students. Therefore, the fourth requirement for a successful engineering education reform is:

Curriculum and pedagogy are revised to make engineering education more likely to attract, retain, and graduate qualified students into the profession, without compromise to quality or content.

Program-level scope of the reform effort. Many dedicated engineering educators have responded to the needs for reform of engineering education, and many in industry, government, and accrediting bodies have tried to help. These efforts can be characterized by their nature and scale: 1) small scale at the level of a course or module; 2) program scale at the level of a degree program, 3) consortia of universities or programs working together; and, 4) research programs on education.

In any program, there are faculty who are exceptionally dedicated to teaching. Universities and funding sources often invest resources in these faculty members to develop new pedagogical approaches based on practice and new content. These faculty members often receive departmental and university awards for teaching and are revered by their students. They are important sources of new ideas and form a pool of early adopters in systemic reform efforts. However, an individual faculty member cannot easily influence an entire program. The reform of engineering education must be addressed on a department or degree program level at the very least. In this way, common expectations for faculty performance and student responsibility for learning can be set and maintained. The educational program must not be viewed as a set of elements, but as a system in which each element carries both individual and collective learning objects for the program. Thus, the fifth requirement for success in engineering education reform:

Any successful attempt at engineering education reform includes most, or all, of the learning experiences from which a student benefits, and, therefore, must be set and maintained at a program or department level.

Collaboration for engineering education reform. A number of university consortia around the world are working on engineering education reform. (See Table 2.1) For example, the IDEA League is an international consortium of four major research universities in London, Delft, Aachen, and Zürich. There are many advantages to working with university consortia when they are properly structured—the principal being acceleration of effort. Consider, for example, a reasonable timeline for systemic education reform: in Year 1, an opportunity for improvement is identified, and an approach developed; in Year 2, the approach is tested; in Year 3, it is refined and re-tested; and, in Year 4, it is arguably finalized. Now consider the tasks associated with this reform: a) the curriculum—what will be taught and where; b) the pedagogical

TABLE 2.1. EXAMPLES OF UNIVERSITY CONSORTIA FOR ENGINEERING EDUCATION REFORM

Group name/location	Affiliated institutions	Focus and projects
The IDEA League http://www.idealeague.org	Imperial College London TU Delft RWTH Aachen ETH Zürich	Exchange of ideas and expertise in education and research in science and technology
Center for the Advancement of Engineering Education (CAEE) http://www.engr.washington.edu/caee	Colorado School of Mines Howard University Stanford University University of Minnesota University of Washington	Collaboration focused on scholarship on learning engineering, teaching engineering, and engineering education. Share publications, presentations, and workshops.
Center for the Integration of Research, Teaching, and Learning (CIRTL) http://cirtl.wceruw.org	University of Wisconsin Michigan State University Pennsylvania State University	Development of a national faculty in science, technology, engineering, and mathematics (STEM). Graduate-through-faculty development founded on teaching-as-research concepts implemented within learning communities.
National Center for Engineering and Technology Education (NCETE) http://www.ncete.org	University of Georgia University of Illinois University of Minnesota Utah State University	One of 17 NSF-funded Centers for Learning and Teaching. Community of researchers in engineering and technology education.Goal is to infuse engineering skills into K-12 schools.
VaNTH Center for Bioengineering Educational Technologies http://www.vanth.org	Vanderbuilt University Northwestern University University of Texas Harvard-MIT	Exchange of research in bioengineering, learning science, learning technology, and assessment.

component—how the curriculum will be taught; c) the evaluation component—how the intended outcomes will be measured and improved; and d) workspace and logistics—the learning environment. The advantages of a consortium are parallel development and shared tasks. As a team, collaborating universities identify common opportunities for improvement, implement several different approaches simultaneously, and compare results based on common evaluation tools. This collaboration greatly accelerates reform efforts. It also allows the sharing of resources and experience, which reduces the cost of transition and increases the likelihood of success. These benefits can be summarized as a sixth requirement for successful educational reform:

Engineering education reform is undertaken by a consortium of programs or departments to allow parallel development and the sharing of resources.

Founded on best-practice educational approaches. Likewise, there are a number of engineering education reform efforts around the world that are research based, that is, they seek to identify best practice and to develop new approaches based on learning theory. For example, the National Academy of Engineering in the United States coordinates a number of research centers and projects through its Center for the Advancement of Scholarship on Engineering Education (CASEE) [6]. Engineering faculty are seldom aware of educational theories and practices that could help them accelerate reform efforts. Many of these research-based initiatives have been successful at bringing together interested parties from both engineering and education to build stronger teams. Some research centers focus on one specific technical discipline, for example, biomedical engineering. Others are more broadly applicable. This leads to the seventh requirement for successful engineering education reform:

Engineering education reform is built on a well-informed adoption of best practice and understanding of models of learning that are broadly applicable to engineering disciplines.

Not demanding of significant new resources. All academic programs exist within an environment of limited resources. This is true across the range of institutions, including polytechnical and research intensive universities. When entering into a program of education reform, we must differentiate between resources needed in the transition and resources in steady state. It is inevitable that in the reform transition, some extra resources, supplied by the teaching staff themselves or preferably by the university, will be needed. Change is not without cost. However, in steady state, we cannot expect more resources, and, therefore, must find new approaches that largely retask existing resources—faculty time, student time, space, etc. This leads to the eighth and final requirement for successful engineering education reform:

Engineering education reform is based on retasking existing resources during ongoing operation.

The CDIO Initiative was designed and developed to meet these eight requirements.

THE CDIO INITIATIVE

The CDIO Initiative is an approach to the contemporary reform of engineering education. It strives to meet the eight requirements for successful engineering education reform, as defined in the previous section of this chapter. It is founded on three key ideas: a set of *goals*, a *vision* or concept for engineering education, and a *pedagogical foundation* that ensures that the vision is realized. These three key ideas are presented in sequence in this section.

The goals

The CDIO Initiative has three overall goals: *To educate students who are able to:*

1. *Master a deeper working knowledge of technical fundamentals*
2. *Lead in the creation and operation of new products, processes, and systems*
3. *Understand the importance and strategic impact of research and technological development on society*

For the reasons discussed above, we believe that these three goals are best met by making the context of the engineering education one of conceiving, designing, implementing and operating. Let's begin by discussing the goals in some detail.

Goal #1. Engineering education should always emphasize the technical fundamentals. University is the place where the foundations of subsequent learning are laid. Nothing in our approach is meant to diminish the importance of the fundamentals, or students' need to learn them. In fact, deep working knowledge and conceptual understanding is emphasized to strengthen the learning of technical fundamentals.

Conceptual understanding is the ability to apply knowledge across a variety of unencountered instances or circumstances [7]. It is not memorization of facts and definitions, nor is it the simple application of a principle that contains the concept, for example, applying the First Law of Thermodynamics. Rather, conceptual understanding represents ideas that have lasting value and offers the potential to engage students. Traditional teaching uses a transmittal approach in which students are assumed to gain knowledge while passively listening to lectures. In a CDIO program, the goal is to engage students in constructing their own knowledge, confronting their own misconceptions. The transition to conceptual-change instruction from the long-standing transmittal approach is difficult. Marton and Säljö [8] call this transmittal approach *a surface approach to learning*, and contrast it with *a deep approach to learning*. Table 2.2 is an adaptation of Marton and Säljö's seminal work, based on the writings of Gibbs [9], Rhem [10], and Biggs [11]. The statement of the goal of educating students who are able to *master a deeper working knowledge of the technical fundamentals* is meant to contrast this approach with that of the more prevalent transmittal approach in current practice. This idea is addressed again in Chapter Six.

TABLE 2.2. **A SURFACE APPROACH TO LEARNING VS. A DEEP APPROACH TO LEARNING**

A surface approach is encouraged by	A deep approach is encouraged by
Excessive amount of material in the curriculum	Student perception that deep learning is required
Relatively high class contact hours	Motivational context
Lack of opportunity to pursue subjects in depth	Well-structured knowledge base
Lack of choice over subject and methods of study	Learner activity and choices
Threatening and anxiety-provoking assessment	Assessment based on application to new situations
Competitive environment	Interaction with others and collaboration

Goal #2. The second goal is to educate students who are able to *lead in the creation and operation of new product, processes, and systems.* This goal recognizes the need to prepare students for a career in engineering. The need to create and operate new products, processes, and systems drives the educational goals related to personal and interpersonal skills; and product, process, and system building skills. Personal skills and attitudes include modes of thought, for example, engineering reasoning and problems solving, scientific inquiry, system thinking, and critical and creative thinking. Personal attitudes and attributes include integrity, accountability, curiosity, a propensity to take risks, and flexibility. Interpersonal skills encompass communication and teamwork. Product, process, and system building skills and knowledge lay the foundation of conceiving, designing, implementing, and operating products and systems within an enterprise and societal context. The more specific learning outcomes that flow from this goal are discussed in a later section and are the main focus of Chapter Three.

Goal #3. The third goal is to educate students who are able to *understand the importance and strategic impact of research and technological development on society*. Our societies rely heavily on the contributions of scientists and engineers to solve problems, ranging from healthcare to entertainment, and to ensure the competitiveness of nations. However, research and technological development must be paired with social responsibility and a move toward sustainable technologies. Graduating engineers must have insight into the role of science and technology in society to assume these responsibilities. This goal further recognizes that some students will not become practicing engineers, but will pursue careers as researchers in industry, government, and higher education. Despite different career interests, all students benefit from an education set in the context of product, process, and system development. First, they benefit from fulfillment of the first goal of deep learning of technical fundamentals. Second, engineering researchers need to understand the connection between their efforts and the eventual impact on a product or system. Successful researchers are increasingly identified for their impact on society in addition to their scholarship. Therefore, it is important for students who embark on careers in research to understand how technology infuses products and processes, and be able to judge and improve the strategic value of their work.

The first two goals represent the historic and contemporary tension in engineering education, *that is,* between knowledge of technical fundamentals and skills. Most engineering educators agree that these two goals are important, but they disagree about how much time to spend on one versus the other. If the model of education is a transmittal process with fixed maximum effective transmittal rate and fixed duration, the tension between technical fundamentals and skills intensifies. The CDIO Initiative has an alternate view of education that helps to relieve that tension. We assert that it is possible to strengthen the learning of the fundamentals and at the same time, improve the learning of personal, interpersonal skills and product, process and system building skills.

The third goal represents another significant tension in engineering education, that between research and development. With an integrated curriculum, it is possible to address both areas, and give students the option of where to place more emphasis in their own career preparation.

The vision

In order to resolve this tension, a new vision for engineering education is needed. This education needs to be based on a scholarship of learning and on best practices of engineering education. It should be integrated and comprehensive, that is, encompassing the entire educational program.

The CDIO Initiative envisions an education that stresses the fundamentals, set in the context of Conceiving—Designing—Implementing—Operating products, processes, and systems. The salient features of the vision are that:

- Education is based on clearly articulated program goals and student learning outcomes, set through stakeholder involvement.
- Learning outcomes are met by constructing a sequence of integrated learning experiences, some of which are experiential, that is, they expose students to the experiences that engineers will encounter in their profession.
- Proper crafting of these integrated learning experiences will cause them to have *dual impact*, simultaneously teaching skills and supporting the deeper learning of fundamentals.

In the discussion below, we present our approach to setting learning outcomes, followed by brief overviews of key aspects of the CDIO vision:

- A curriculum organized around mutually supporting disciplinary courses with activities highly interwoven that develop personal and interpersonal skills, and product, process and system building skills.
- Design-implement and hands-on learning experiences set in both the classroom and in modern learning workspaces as the basis for engineering-based experiential learning.
- Active and experiential learning, beyond design-implement experiences, that can be incorporated into disciplinary courses.
- A comprehensive assessment and evaluation process.

We must find ways to realize this vision by strengthening the collective competence of the faculty, by retasking existing resources, and by not expecting substantially more new resources.

Learning outcomes. The first task of turning the vision into a model program was to develop and codify a comprehensive understanding of abilities needed by contemporary engineers. This task was accomplished through the use of stakeholder focus groups comprised of engineering faculty, students, industry representatives, university review committees, alumni, and senior academicians. The focus groups were asked the first of the two central questions that must be addressed in the reform of engineering education, "*What is the full set of knowledge, skills, and attitudes that engineering students should possess as they leave university?*" An example of thoughtful input from industry received through this process is that of Ray Leopold, former Vice President and Chief Technology Officer of Motorola's Global Telecom Solutions Sector. (See Box 2.3). Results of the focus groups, plus topics extracted from the views of industry, government,

Box 2.3. The Need for CDIO Engineers in Industry

In my estimation, the greatest potential contribution of graduates of CDIO programs is their ability to perform their engineering skills with a more mature appreciation of how a product satisfies real societal needs. This requires project success, broadly defined, which is based on both engineering, and non-engineering contributions.

The engineer must be able to find not only engineering solutions to a problem, but also economic solutions that have a high potential of being successful. The engineer must define value propositions and find solutions to them. A graduating student must develop the skills not only to create brilliant new ideas, but also to transform those ideas into new realities.

As part of this process, engineering graduates must have a better understanding of the value they add to the organization. They must have better developed personal skills, and be able to work with other engineers and with colleagues from other disciplines. The maturity of an engineer flows not only from knowledge of the breadth and depth of disciplinary knowledge, but also from the individual's experience in developing personal and professional skills.

Within industry, we generally try to determine what an individual knows, how an individual can contribute, the perspective an individual brings to us, and how well the individual fits into the culture of our organization. We often do not hire high-powered technologists who don't exhibit the people skills to fit into our team environment, or whose perspective seems to be limited to a narrow technical field. We want deep technical expertise, but that expertise must have a context, and the individual needs to be able to work with others.

In an interview, I often ask behaviorally oriented questions, such as, "*From your educational experiences, tell me specifically about a time when you had to:*

- *deal with a person who didn't seem to be focused on the team goals*
- *redefine a value proposition*
- *adjust your work plans to meet a schedule.*

The graduate of a CDIO program should be able to respond more richly to these questions, and their responses should connote an appreciation for the bigger picture while satisfying the problem at hand.

–R. Leopold, The Motorola Corporation

TABLE 2.3. THE **CDIO** SYLLABUS AT THE SECOND LEVEL OF DETAIL

1 TECHNICAL KNOWLEDGE AND REASONING	**3 INTERPERSONAL SKILLS: TEAMWORK AND COMMUNICATION**
1.1 KNOWLEDGE OF UNDERLYING SCIENCE	3.1 MULTI-DISCIPLINARY TEAMWORK
1.1 CORE ENGINEERING FUNDAMENTAL KNOWLEDGE	3.2 COMMUNICATIONS
1.2 ADVANCED ENGINEERING FUNDAMENTAL KNOWLEDGE	3.3 COMMUNICATIONS IN FOREIGN LANGUAGES
2 PERSONAL AND PROFESSIONAL SKILLS AND ATTRIBUTES	**4 CONCEIVING, DESIGNING, IMPLEMENTING, AND OPERATING SYSTEMS IN THE ENTERPRISE AND SOCIETAL CONTEXT**
2.1 ENGINEERING REASONING AND PROBLEM SOLVING	4.1 EXTERNAL AND SOCIETAL CONTEXT
2.2 EXPERIMENTATION AND KNOWLEDGE DISCOVERY	4.2 ENTERPRISE AND BUSINESS CONTEXT
2.3 SYSTEM THINKING	4.3 CONCEIVING AND ENGINEERING SYSTEMS
2.4 PERSONAL SKILLS AND ATTITUDES	4.4 DESIGNING
2.5 PROFESSIONAL SKILLS AND ATTITUDES	4.5 IMPLEMENTING
	4.6 OPERATING

and academia on the expectations of university graduates were organized into a list of learning outcomes, called the CDIO Syllabus. The description, development, and validation of the Syllabus are the subjects of Chapter Three.

As shown in Table 2.3, the CDIO Syllabus classifies learning outcomes into four high-level categories:

1. Technical knowledge and reasoning
2. Personal and professional skills and attributes
3. Interpersonal skills: teamwork and communication
4. Conceiving, designing, implementing, and operating systems in the enterprise and societal context

These four headings map directly to the underlying need identified in an earlier section of this chapter, that is, to educate students who can:

> *understand how to conceive, design, implement, and operate (Section 4)*
> *complex value-added engineering products, processes, and systems (Section 1)*
> *in a modern team-based engineering environment (Section 3), and*
> *are mature and thoughtful individuals (Section 2).*

The last phrase, "are mature and thoughtful individuals", acknowledges that within a university context, students grow psychologically and socially, as well as intellectually. The knowledge, skills and attitudes outlined in Sections 2, 3 and 4 of the Syllabus are referred to as *personal and interpersonal skills; and product, process, and system building skills*. The first section, *Technical Knowledge and Reasoning*, is program-specific, that is, it outlines major concepts of a specific engineering discipline. Sections 2, 3, and 4 are applicable to any engineering program.

The content of each section was expanded to second, third and fourth levels. Syllabus topics at the second level of detail were validated with subject experts and key stakeholders. To ensure comprehensiveness, the Syllabus was explicitly correlated with documents listing engineering education requirements and desired attributes. As a result, the CDIO Syllabus is a rational and consistent set of skills, derived from an understanding of needs, that stakeholders would expect from graduating students. It is comprehensive, peer reviewed, and forms the basis for program design and assessment. Table 2.3 is the CDIO Syllabus at the second level of detail. The complete Syllabus is found in Appendix A.

To translate the CDIO Syllabus topics and skills into assessable learning outcomes, we proposed a survey of program stakeholders to determine the level of proficiency expected of graduating engineers in each of the Syllabus topics. The survey process and results from representative programs are explained in Chapter Three. As a result of the initial surveys, we have a stakeholder-based, comprehensive answer to the first of the two central questions posed at the beginning of this chapter, "*What is the full set of knowledge skills and attitudes that engineering students should possess as they leave the university, and at what level of proficiency?*" The remaining features of the CDIO vision address the second central question, "*How can we do better at ensuring that students learn these skills?*" Broadly speaking, this requires reforms in four major areas: the structure of the curriculum of our programs and the content of some our courses; the learning environment in which we teach; the way we teach; and they way in which we assess and evaluate the outcomes—on the student level as well as on the program level. Our approaches in these four areas are based on educational research we have conducted, and on broad surveys of best practice. Then it further evolved through review by collaborators. A complementary view, which outlines desired progress in similar areas is included in Box 2.4, written by Professor Sheri Sheppard of Stanford University and her colleagues at The Carnegie Foundation for the Advancement of Teaching.

Curriculum reform. To achieve the dual goals of deeper working knowledge of technical fundamentals and ability to lead in the creation and operation of new products, processes, and systems, we must improve the engineering curriculum. We cannot expect more resources, longer terms, more years, or other extensions to the curriculum. Consequently, we must retask existing resources. The challenge is to develop an *integrated curriculum*, that is, to find innovative ways to make double duty of teaching time so that students develop a deeper working knowledge of technical fundamentals while simultaneously learning personal, and interpersonal skills; and product, process, and system building skills.

We should not leave this learning to chance, but have an explicit plan for ensuring that students learn these skills. Accomplishing this integration may require changes to curriculum structure that exploit extra- and co-curricular

Box 2.4. Educating Engineers (Based on a Carnegie Study of Engineering Education in the U. S. [1])

Formal engineering education is responsible for students learning the skills necessary to embark on successful engineering careers and to contribute to the engineering needs of the nation. From the students' point of view, entrance into engineering school is the beginning of a three-part apprenticeship—a cognitive or intellectual apprenticeship, a practical apprenticeship of skill, and an apprenticeship of professional identity and values. This is also the case for those entering other professions, such as law or medicine.

Curriculum

Three of the major components of formal engineering education are the curriculum, the pedagogies employed, and the program. The *curriculum* should reflect the skills, knowledge, practices, and values found in engineering work. As such, it should include coverage of core knowledge, key problem solving strategies, and the use of knowledge to resolve new, novel, and/or significant problems.

Core knowledge in engineering work includes: theoretical tools (math-based and conceptual); fundamental design concepts (operational principles and normal configurations); criteria and specifications; quantitative data; practical considerations; process-facilitating tools; and contextual knowledge. Formal education should focus on the knowledge-types with the broadest and most enduring value to support continued professional learning and practice, including theoretical knowledge, operational principles, process-facilitating tools, and contextual knowledge [2,3].

Key problem solving strategies involve design and analysis, where a problem or current state is identified, attributes and constraints are defined, and a means-end relationship is developed [3,4]. Finally, the curriculum should include guided experiences in practice, defined broadly as the minimally scripted application of knowledge and problem solving strategies to responsibly resolve an ill-defined problem.

Pedagogies

The *pedagogies* used to deliver the curriculum need to be selected with care. They should only be chosen after an analysis of learning goals and consideration of how students' progress in achieving these goals will be assessed [5]. Successful implementation of any pedagogy also requires the teacher to have "pedagogical content knowledge" that goes well beyond the content knowledge of a discipline [6]. The teaching methods should support knowledge transfer and development toward expertise [7]. Emerging research findings in the area of learning and cognition, especially research on perceptual learning, the development of expertise in a variety of fields, and problem solving, suggest new ways to review and improve current teaching models in engineering.

Program

The *program* represents the third formal component of engineering education. The program is largely composed of teachers bringing the curriculum to students via various pedagogies. In order to promote students' progress toward acquiring expertise in engineering, a program should not simply be a collection of courses. Rather, it must be designed and delivered as a suite of interacting and interlocking educational experiences focused by the goal of developing students' abilities to carry out engineering tasks. In thinking about a particular program, its faculty should engage actively in discussion, debate and action as a community around such questions as:

- What types of knowledge are included? Why are these types of knowledge emphasized? Who decides this? What kinds of teaching, learning, and assessment are employed to ensure that students are actually learning these types of knowledge?
- In what ways is engineering problem solving taught? What methods are used to teach analytic problem solving? What methods are used for teaching synthetic problem solving? For integrated problem solving? How might the methods be enhanced to more tightly draw in core knowledge?
- What types of educational experiences challenge students to integrate specialized knowledge and problem solving? In other words, what pedagogies lead students to engage in the actual practices of engineering?

- In what ways are students challenged to integrate contextual information and knowledge into problem solving? How are students taught to act with professionalism? How might things be improved?
- What is the relationship between practice and education? What should this relationship be? Who should be involved in defining this relationship? How does history influence this relationship?
- Does the form (number and progression of courses) in the program represent a reasonable four-year workload? Who should have a stake in defining what is reasonable and/or optimum? What should be the balance between knowledge-learning and problem-solving practice?
- Are the educational practices in our program up to the task of educating future professionals? How will we know that we are successful in this task?

References

[1] Sheppard, S., Sullivan, W., and Colby, A., "Preparation for the Professions Program: Engineering Education in the United States," in *Educating the Engineer of 2020: Adapting Engineering Education to the New Century*, 2005. Available at http://www.nap.edu/books/0309096499/html/

[2] Vincenti, W. G., *What Engineers Know and How They Know It: Analytical Studies from Aeronautical History*, The Johns Hopkins University Press, Baltimore, Maryland, 1990.

[3] Sheppard, S., Colby, A., Macatangay, K., and Sullivan, W., "What is Engineering Practice?", in press for the *International Journal of Engineering Education.*

[4] Rubinstein, M. F., *Patterns of Problem Solving*, Prentice-Hall, Englewood Cliffs, New Jersey, 1975.

[5] Bransford, J., Vye, N., Bateman, H., "Chapter 6—Creating High-Quality Learning Environments: Guidelines from Research on How People Learn", in *The Knowledge Economy and Postsecondary Education: Report of a Workshop*. Available at http://books.nap.edu/catalog/10239.html

[6] Shulman, L. S., "Knowledge and Teaching: Foundations of the New Reform", *Harvard Educational Review*, 57, 1-22, 1987.

[7] Schwartz, D. L., Bransford, J. D., and Sears, D. L., "Efficiency and Innovation in Transfer", in Mestre, J. (Ed.), *Transfer of Learning From a Multidisciplinary Perspective*, Information Age Publishing.

– S. Sheppard, W. Sullivan, A. Colby, K. Macatangay,
The Carnegie Foundation for the Advancement of Teaching

and extra-campus learning opportunities, and the development of new teaching materials. To facilitate curriculum reform, we suggest retaining the disciplinary courses as the organizing structure of the curriculum, while making two substantive improvements. First, the disciplinary courses must work together to be mutually supporting, as they are in practice. Second, education in personal and interpersonal, and product, process and system building skills must be interwoven into the disciplinary education.

Designing a new curriculum requires benchmarking of the current curriculum to identify existing connections among disciplines and places where skills are already taught, and to identify omissions and overlaps. Three specific curricular structures are key elements of an integrated curriculum: 1) an *introductory engineering experience* that creates the framework for subsequent learning and motivates students to be engineers; 2) conventional disciplinary courses coordinated and linked to demonstrate that engineering requires interdisciplinary efforts; and, 3) a final project course—or capstone—that includes a substantial experience in which students conceive, design, implement, and operate a product, process, or system. With these new structures in place, an explicit plan to overlay skills can be developed. The new curriculum structure also facilitates co-curricular student projects, internships, and placements in industry that can significantly expand the time available for learning skills and enrich the overall learning experience. The result of such curricular reform is an integrated curriculum, which contains a sequence of well-planned learning experiences that help students meet the rigorous and appropriate educational goals. Chapter Four describes the design and development of an integrated curriculum.

Design-implement experiences and CDIO workspaces. Engineers design and implement products, processes, and systems. Providing students with repeated *design-implement experiences* helps them develop deep working knowledge of the fundamentals and learn the skills to design and implement new products, processes, and systems. Since personal and interpersonal, and product, process and system building skills are derived from engineers' need to work in design teams, design-implement projects provide a natural setting in which to teach students these skills. In a CDIO program, experiences in conceiving, designing, implementing, and operating are woven into the curriculum, particularly in the introductory and concluding project courses. The concluding project course can be retasked into one that is closely linked to one or more disciplines and engages students in designing, implementing, and operating a product, process, or system. Aligning theory development with practical implementation gives students opportunities to learn both the applicability and limitations of theory.

If students are to understand that conceiving—designing—implementing—operating is the context of the education, then it is desirable to retask existing laboratory space by building modern engineering workspaces that are supportive of, and, organized around C, D, I, and O. In such *CDIO workspaces*, the *Conceive* spaces are designed to encourage people to interact and to understand the needs of others and to provide a venue which encourages reflection and conceptual development. They are largely technology-free zones. *Design* and *Implement* facilities introduce students to digitally enhanced collaborative design and modern fabrication and integration of hardware and software. *Operate* workspaces are more difficult to manage in academic settings. However, students can learn how to operate their own and faculty-assigned experiments. Simulations of real operations, as well as electronic links to real operations environments can supplement the direct student experience. In addition, workspaces must also support other modes of active and hands-on learning, including experimentation, disciplinary laboratories, and social interaction. The space must facilitate and encourage team building and team activities. Design-implement experiences and CDIO workspaces are explored in Chapter Five.

Teaching and learning reform. Having addressed curriculum issues of what to teach, we now consider the pedagogical issues of how to teach and how students learn. To meet the dual goals of improved disciplinary learning and skills learning, it is necessary to retask students' learning time and employ best practices in teaching and learning throughout the program. To address these learning needs, we recommend improvement in two basic areas: 1) an increase in *active and experiential learning*, and 2) the creation of *integrated learning experiences* that lead to the acquisition of both disciplinary knowledge, personal and interpersonal skills; and product, process, and system building skills.

Educational research confirms that active learning techniques significantly increase student learning. Active learning occurs when students are more involved in manipulating, applying, and evaluating ideas. Active learning in lecture-based courses can include pauses for reflection, small group discussion, and real-time feedback from students about what they are learning. Active learning becomes experiential when students take on roles that simulate professional engineering practice, that is, design-implement projects, simulations, and case studies. The emphasis on widespread use of active and experiential learning is a major aspect of our commitment to develop deeper working knowledge of the technical fundamentals. The desired outcome is an understanding of the underlying technical concepts, as well as their application. This is understood to be a precursor to innovation.

To make more effective and efficient use of student learning time, integrated learning experiences are required. Integrated learning refers to learning experiences that lead to the acquisition of disciplinary knowledge concurrently with personal and interpersonal skills, and product, process, and system building skills. This gives the learning experiences *dual impact*. This learning certainly occurs in design-implement experiences, but is not limited to these experiences. For example, solving problems is an essential skill of engineering. Disciplinary knowledge allows a student *to solve the problem right,* but an integration of broader skills is necessary to teach students *to solve the right problem.* The CDIO approach aims to develop skills in problem formulation, estimation, modeling and solution. A modified problem-based learning format, with strong emphasis on the fundamentals, supports this type of integrated learning. However, there are many other opportunities to integrate learning, for example, coupling communication or teamwork with an assignment; encouraging students to dig deeply into a topic and use specific research and inquiry methods; or, discussing the ethical aspects of a technical problem concurrently with its technical aspects. An important subtle aspect of this integrated learning is that students see their role models, namely, the engineering faculty, discussing this wider range of skills, signaling their importance to the profession. Integrated learning and active and experiential learning are the focus of Chapter Six.

Assessment and evaluation. Rigorous assessment and evaluation are required to guide the educational reform process. The *learning assessment* component measures student learning and monitors achievement of disciplinary, personal, interpersonal, product, process, and system building learning outcomes. The *program evaluation* component gathers and analyzes data related to the overall quality and impact of the entire educational program.

Effective learning assessment focuses on the intended outcomes for students, that is, the knowledge, skills, and attitudes that students are expected to master as a result of their educational experiences. Student learning assessment measures the extent to which each student achieves specified learning outcomes. Learning assessment methods include written and oral exams, observation and rating of oral presentations and other processes, peer assessment, self-assessment, and portfolios. In a CDIO program, assessment is learner-centered: it is aligned with teaching and learning outcomes, uses multiple methods to gather evidence of achievement, and promotes learning in a supportive, collaborative environment. Assessment focuses on gathering evidence that students have developed proficiency in disciplinary knowledge, and personal and interpersonal skills; and product, process, and system building skills. Learning assessment is the focus of Chapter Seven.

Program evaluation is a judgment of the overall quality of a program based on evidence of a program's progress toward attaining its goals. Data collection techniques include best-practice methods of program evaluation, such as entry interviews, student satisfaction surveys, and instructor reflective memos. When evidence and results are regularly reported back to faculty, students, program administrators, alumni, and other key stakeholders, the feedback becomes the basis for making decisions about the program and its continuous improvement. Program evaluation and continuous improvement are discussed in Chapter Nine.

Pedagogical foundation

We believe that reforming engineering education based on the CDIO vision will bring us closer to resolving the tension between the two primary goals of developing deeper learning of the technical fundamentals and the ability to lead in the creation and operation of products, processes, and systems. This belief is based not only on experience, but also on application of theories and models of learning.

To understand pedagogical improvements, we must consider what we know about how students learn. As is the case with most children and adults, many engineering students tend to learn from the concrete to the abstract. Yet, they no longer arrive at universities armed with hands-on experiences from tinkering with cars or building radios. Likewise, the engineering science educational reforms of the latter half of the 20th century largely removed many of the hands-on experiences that engineering students once encountered at university. As a result, contemporary engineering students have little concrete experience upon which to base engineering theories. This lack of practical experience affects students' ability to learn abstract theory that forms much of the engineering fundamentals, and also hampers their ability to realize the applicability and practical usefulness of a good theory.

The CDIO approach is based on experiential learning theory that has roots in constructivism and cognitive development theory. Cognitive development theorists, among whom Jean Piagét is perhaps the most influential [12], explain that learning takes place in developmental stages. The ideas of Piagét and cognitive development theorists who followed him, led to three important principles about learning that bear on our programs:

- The essence of learning is that it involves teaching learners to apply cognitive structures they have already developed to new content.
- Because learners cannot learn to apply cognitive structures they do not yet possess, the basic cognitive architecture must first evolve on its own.
- Learning experiences that are designed to teach concepts that are clearly beyond the current stage of cognitive development are a waste of time for both teacher and learner [13].

Cognitive development theories, in conjunction with social psychology and social learning theory, provide historical precedents for constructivism, a theory that postulates that what is learned is a function of the content, context, activity, and goals of the learner. Constructivists believe that learners build their internal frameworks of knowledge upon which they attach new ideas. Individuals learn by actively constructing their own knowledge, testing concepts on prior experience, applying these concepts to new situations, and integrating the new concepts into prior knowledge. Facilitating the processing of new information and helping students to construct meaningful connections is regarded as the basic requirement for teaching and learning.

The theories of constructivism and social learning have been applied to a number of curriculum and instruction models and practices. The CDIO approach focuses on one of these practices, called experiential learning. Experiential learning can be defined as the process of creating and transforming experience into knowledge, skills, attitudes, values, emotions, beliefs and senses. In his work on experiential learning, Kolb [14] emphasizes six characteristics of experiential learning:

- Learning is best conceived of as a process, that is, concepts are derived from, and continuously modified by, experience.
- Learning is a continuous process grounded in experience, that is, learners enter the learning situation with more or less articulate ideas about the topic at hand, some of which may be misconceptions.
- The process of learning requires the resolution of conflicts between opposing modes of adaptation to the world, that is, the learner needs different abilities from concrete experience to abstract conceptualization, and from reflective observation to active experimentation.
- Learning is a holistic process of adaptation to the world, that is, learning is broader than what occurs in classrooms.
- Learning involves transactions between the person and the real-world environment.
- Learning is a process of creating knowledge, that is, in the tradition of constructivist theories.

In this light, the essential feature of the CDIO approach—that it creates *dual-impact* learning experiences—can be better understood. If the experiential learning activities are crafted to support explicit pre-professional behavior, they will facilitate the learning of personal and interpersonal skills, and of product, process and system building skills. More subtly, these learning experiences allow the student to develop a knowledge structure for understanding and learning the abstractions associated with the technical fundamentals. The concrete experiences also provide opportunities for active application that supports understanding and retention. Thus, they provide the pathway to the desired goal—deeper working knowledge of the fundamentals.

Meeting the requirements

In this discussion, we have demonstrated that the CDIO approach meets four of the eight requirements for successful engineering education reform:

- It stresses the technical fundamentals and improves learning of personal and interpersonal skills, and product, process, and system building skills.
- The product, process, and system lifecycle is the context of the education.
- Educational goals and learning outcomes reflect the input of all stakeholder groups.
- It is based on well-informed best practice and understanding of learning models that are broadly applicable to engineering disciplines.

In the next section of this chapter, we will show how a CDIO approach meets the remaining four requirements for successful reform.

REALIZING THE VISION

As described earlier in this chapter, the CDIO Initiative addresses the widely recognized *need* to educate students who understand how to conceive-design-implement-operate complex value-added engineering products, processes, and systems in a modern team-based environment. The key program *goals* are to educate students who can master a deeper working knowledge of technical fundamentals, lead in the creation and operation of new products, processes, and systems, and understand the importance and strategic impact of research and technological development on society. We believe these goals are reached when conceiving-designing-implementing-operating products, processes, and systems is the *context* of the education. The *vision* includes learning outcomes set through stakeholder engagement, and an education centered on a sequence of integrated experiential learning experiences, set in a curriculum organized around mutually supporting technical disciplinary courses with personal and interpersonal skills; and product, process, and system building skills highly interwoven. The *pedagogical foundation* supports the premise that with well-planned concrete experiences in engineering, coupled with active and experiential learning, the goals can be reached with existing resources.

The challenge in realizing the vision is to transform engineering programs and, in fact, the culture of engineering education. To aid in this transformation, we have adopted a number of techniques to engage faculty, facilitate progress, and ensure quality:

- A rigorous statement of goals for student learning, that is, the CDIO Syllabus
- A clear set of programmatic characteristics that distinguish a CDIO program, that is, the CDIO Standards
- Support for organizational and cultural change
- Enhancement of faculty competence in both skills and in teaching, learning, and assessment
- Shared open-source resources so that, in the steady state, a reformed program is not substantially more resource intensive than a standard program
- Collaboration of programs for parallel development and approaches to common issues
- Alignment with national standards and other major reform initiatives

The immediate outcome of the CDIO approach is to attract and interest students and to educate engineers who are "ready to engineer." Each of these techniques is described briefly here, and explained in more detail in subsequent chapters. The first two—the CDIO Syllabus and CDIO Standards—constitute the *What* and *How* of educational reform, as suggested by Figure 2.3.

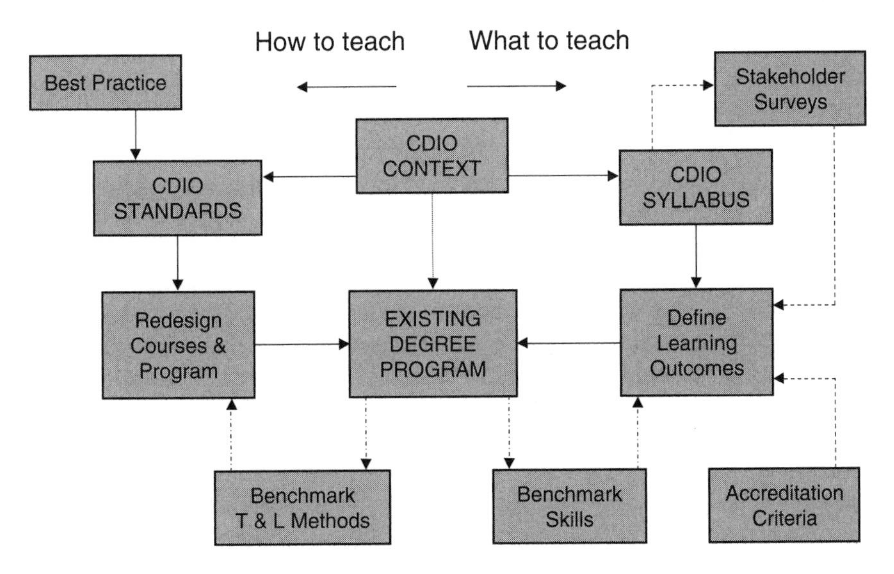

FIGURE 2.3. IMPLEMENTING THE CDIO APPROACH

The CDIO Syllabus

The starting point for educational design and development is the statement of learning outcomes, that is, the capabilities or competencies that students should possess upon completion of a course or program. This statement of learning outcomes is the answer to the first central question, *"What is the full set of knowledge, skills, and attitudes that engineering students should possess as they leave the university, and at what level of proficiency?"* Clear statements of learning outcomes play a key role in educational design by

- Formalizing the knowledge, skills, and attitudes that alumni, faculty, industry leaders and society expect from engineering graduates
- Supporting the design of an integrated curriculum (see Chapter Four), integrated learning experiences (see Chapter Six), and systematic assessment of student learning (see Chapter Seven)
- Providing information for current and future students about the program

The CDIO Syllabus, discussed briefly in this chapter, is explained in detail in Chapter Three.

The CDIO Standards

We have developed 12 standards that describe CDIO programs. They codify the guiding principles in designing and developing a program. They are the outline of the answer to the second central question, *"How can we do better at ensuring that students learn these skills?"* The Standards were developed in response to program leaders, alumni, and industrial partners who wanted to know how they would recognize CDIO programs and their graduates. As a result, these standards define the distinguishing features of a CDIO program, serve as guidelines for educational program reform and evaluation, create benchmarks and goals with worldwide application, and provide a framework for continuous improvement.

The 12 CDIO Standards address

- Program philosophy (Standard 1)
- Curriculum development (Standards 2, 3 and 4)
- Design-implement experiences and workspaces (Standards 5 and 6)
- Methods of teaching and learning (Standards 7 and 8)
- Faculty development (Standards 9 and 10)
- Assessment and evaluation (Standards 11 and 12)

The Standards are also the organizing principle of this book. Each chapter focuses on one or two standards, explaining their meaning and giving examples of their application in existing CDIO programs. Table 2.4 lists the 12 Standards with references to the chapters in which they are discussed. Complete statements of the CDIO Standards are found in Appendix B. For each standard, a description explains the meaning of the standard, highlighting reasons for setting the standard. Examples of evidence the standard is being met is correlated with

Table 2.4. **The CDIO Standards**

	CDIO Standard	Chapter
1	**The Context** Adoption of the principle that product, process, and system lifecycle development and deployment—Conceiving-Designing-Implementing-Operating—are the context for engineering education	2
2	**Learning Outcomes** Specific, detailed learning outcomes for personal and interpersonal skills; and product, process, and system building skills, as well as disciplinary knowledge, consistent with program goals and validated by program stakeholders.	3
3	**Integrated Curriculum** A curriculum designed with mutually supporting disciplinary coursess, with an explicit plan to integrate personal and interpersonal skills; and product, process, and system building skills.	4
4	**Introduction to Engineering** An introductory course that provides the framework for engineering practice in product, process, and system building, and introduces essential personal and interpersonal skills.	4
5	**Design-Implement Experiences** A curriculum that includes two or more design-implement experiences, including one at a basic level and one at an advanced level.	5
6	**Engineering Workspaces** Engineering workspaces and laboratories that support and encourage hands-on learning of product, process, and system building, disciplinary knowledge, and social learning.	5
7	**Integrated Learning Experiences** Integrated learning experiences that lead to the acquisition of disciplinary knowledge, as well as personal and interpersonal skills; and product, process, and system building skills.	6
8	**Active Learning** Teaching and learning based on active and experiential learning methods.	6
9	**Enhancement of Faculty Competence** Actions that enhance faculty competence in personal and interpersonal skills; and product, process, and system building skills.	8
10	**Enhancement of Faculty Teaching Competence** Actions that enhance faculty competence in providing integrated learning experiences, in using active experiential learning methods, and in assessing student learning.	8
11	**Learning Assessment** Assessment of student learning in personal and interpersonal skills; and product, process, and system building skills, as well as in disciplinary knowledge.	7
12	**Program Evaluation** A system that evaluates programs against these standards, and provides feedback to students, faculty, and other stakeholders for the purposes of continuous improvement.	9

each standard. As explained in Chapter Nine, the standards are also used as the basis of program evaluation and continuous improvement.

The focus of this chapter is to develop the context for the CDIO approach and explain the rationale for adopting Conceiving, Designing, Implementing, and Operating as the context of engineering education. This is the intent of Standard 1.

STANDARD 1 – THE CONTEXT

Adoption of the principle that product, process, and system lifecycle development and deployment—Conceiving, Designing, Implementing, and Operating—are the context for engineering education.

The product, process, and system lifecycle is considered the context for engineering education in that it is the cultural framework, or environment, in which technical knowledge and other skills are taught, practiced, and learned. The principle is adopted by a program when there is explicit agreement of faculty to transition to a CDIO program, and support from program leaders to sustain reform initiatives.

Organizational and cultural change

Implementing the CDIO approach implies a fundamental shift in the nature of engineering education to a more integrated curriculum, rich with experiential learning, in the context of product, process, and system building. This will be a challenge. The current faculty of engineering are, by and large, engineering researchers. They tend to think of disciplines in isolation, explain them based on theoretical underpinnings, and focus on the evolution of the discipline, rather than its application or synthesis. Bringing about such a transformation will require more than simply redrafting curricula; it may require cultural change. To be effective in this transformation, we should acknowledge this, and be prepared to learn from best practice in organizational and cultural change, and to apply it accordingly to the university setting. Professor Thomas Gray, of the University of Strathclyde, reflects on this issue of change and implementation in Box 2.5. This will be one of the topics of Chapter Eight.

BOX 2.5. AN IDEAL ENGINEERING EDUCATION

The ideal engineering education cannot be defined. that is, without reference to time, place, environment and context. There may also be some risk in too close a definition, in case it turns into a *tablet of stone*. The thought behind this is that the default position in universities and professional bodies alike tends to be one of resistance to change. Therefore, the starting point for an *ideal* program needs to have many seeds of change sprinkled liberally throughout. Looking back over the years, most progress in my experience has been made through strategies that capitalize on serendipity and permit some anarchy.

With such caveats in place, we can try to identify the skills, capabilities, and attitudes that we would want to develop in graduate engineers, hopefully, not in a boring, legalistic manner. These attributes have always been necessary. In the past, they may not have been addressed directly within education programs, but in the long-distant memory through hobbies, industrial practice, etc. The following points might be said to characterize an effective graduate engineer for current conditions:

- An insatiable curiosity for understanding how things work, over the broadest spectrum of engineering and nature, underpinned by any necessary understanding of hard science or engineering practice

- Sufficient confidence to model and analyze the behavior of engineering systems. This can only be developed through the experience of doing it independently.
- Awareness of, and sensitivity to, the contexts in which engineering is set, that is, social, environmental, economical, and political
- Sufficient confidence to make things happen and artifacts work, based on the experience of doing it against a time and cost budget
- Awareness of the capabilities of self and others
- Ability to make the best of these capabilities, based on the experience of doing and through reflection on how things happened

These attributes have always been important. In the past, however, universities have relied on other areas of the experience of students and graduates to develop contextual awareness, knowledge of engineering practice, and abilities to make things happen. Expectations are now much higher. Graduates are required to hit the ground running, and enterprises range far and wide in terms of scientific and technical scope. This implies more emphasis in engineering programs on learning how to learn, on doing, on self-evaluation, and on collaborating fruitfully with others.

The major challenge in devising and operating education programs, therefore, is to stretch students to develop these characteristics. The major risks lie in reducing everything to compartmentalized exercises; or, at the opposite extreme, having a lot of fun, developing independence, but leaving out professionally critical areas of knowledge, understanding, or capability. The good thing about the active type of program is that it should be substantially more motivating for the student.

A CDIO program is very likely to succeed in achieving the above aims precisely because it reverses the focus. In what might be called *traditional* programs, discipline and topic-based knowledge, understanding and analytical capability dominate. The integrating glue is assumed to be acquired (how is not quite clear), and the capability to make new things happen independently of what has gone before is exercised only through a small input of individual project work. Typical CDIO activities in a real engineering context, on the other hand, are about doing things, and the knowledge/understanding/analysis base is deemed to be subservient to that. As individual CDIO elements cover the entire spectrum of engineering, coverage should be complete, provided that the appropriate engineering science concentration is selected appropriately to the degree discipline.

Of course, it may be argued that CDIO is not the only framework that may be used effectively to achieve these aims. Accreditation bodies in the United Kingdom have recently formulated a matrix coverage approach, whereby aspiring professional engineers are required to demonstrate various attributes cross-referenced to specific engineering contexts: underpinning science, analysis, design, awareness of economic, social and environmental contexts and engineering practice. This looks neat, but it is left to the program provider to decide on activities designed to develop these attributes and to demonstrate to the accrediting bodies that the assessment process is valid relative to a threshold level. Therein lies the problem. The CDIO framework is, of course, entirely compatible with such an approach, but it has the distinct advantage that the models for student activities are focused on typical engineering tasks.

– T. Gray, University of Strathclyde

Enhancement of faculty competence

Part of the change process will require strengthening the competence of faculty in skills and in active and experiential learning and student assessment. There is little reason to expect a faculty that has been recruited as a cadre of researchers to be proficient in many of the skills of engineering practice. And there is absolutely no reason to expect that these faculty researchers would be able to teach these skills. Therefore, if we are to successfully support student learning, we must develop approaches to enhancing the skills of our faculty in engineering.

Likewise, faculty have, by and large, been educated using pedagogical styles based on information transmission—lectures and the like. If we are to develop a learning-focused education, which relies on active and embedded learning, current faculty must be supported in their personal development and use of these techniques. In both cases, skills and teaching, the transformation will be broader and more effective if there is a well-planned effort to build faculty competence, by bringing individuals with this background to the team and enhancing the competence of the existing team. Enhancement of faculty competence is addressed in Chapter Eight.

Open-source ideas and resources

Nothing in the CDIO Initiative is prescriptive. We have developed resources to help engineering programs resolve the essential conflict in engineering education, that is, time and resources for learning both the disciplinary fundamentals and personal and interpersonal skills; and product, process, and system building skills. These resources are intended to facilitate the rapid adaptation and implementation of the CDIO approach into university programs.

To date, it has been implemented in programs that represent differences in goals, students, financial resources, existing infrastructure, university constraints, governmental legislation, industry needs, and professional societies' certification. To accommodate these differences and to acknowledge that our approach is under ongoing development and adaptation, it is codified and documented as an open source. Resources are descriptive, not prescriptive. An open, accessible architecture for the program materials promotes the dissemination and exchange of ideas and resources. These resources are specifically designed so that university engineering programs can adapt the CDIO approach to their specific needs. Engineering programs can implement the entire approach or choose specific components.

The resources available to engineering programs that wish to adapt and implement the CDIO approach include materials that introduce the model, the CDIO Syllabus, survey tools for investigating stakeholder needs, guidelines for design-implement experiences, support for implementation, start-up advice, and suggested steps for the transition. The transition process and its related tools are addressed in more detail in Chapter Eight. Few university programs enjoy the option of increasing available resources. We have designed the approach so that a CDIO program can be implemented with a retasking of existing resources. However, in the transition, some additional time and support may be needed. Chapter Eight describes resources that help to minimize this transitional effort and maximize the benefits of implementing a CDIO program.

Value of collaboration for parallel development

The collaboration of engineering programs in countries worldwide is a fundamental part of our approach to development. Engineering educators around the world struggle with similar issues, for example, the tension between science-oriented goals and practice-oriented skills. Addressing this tension is a challenge for any engineering education designer. The key to effective educational development is not to make minor trade-offs between these two goals, but rather to create a new model for engineering education that encompasses both. This undertaking is difficult for a single program or department. Moreover, the commonality of these issues around the world suggests that many people are working on similar problems. Forming an international initiative enables us to construct and implement a general and adaptable educational model.

Alignment with national standards and other change initiatives

This is an era of increased attention to educational processes in higher education generally, and specifically for engineering. In some cases, national accreditation standards have been revised to reflect an outcomes-based approach to programs. Examples include ABET EC2000 in the United States [15] and UK-SPEC in the United Kingdom [16]. In other cases, reform of higher education is the result of large-scale regional reform, for example, the Bologna Declaration [17], or the project for the Accreditation of Engineering Programmes and Graduates (EUR-ACE) [18].

We have made every attempt to ensure that the CDIO approach is aligned with these efforts. Chapter Three discusses the comparison of the CDIO Standards with several national accreditation standards. These comparisons show a similar trend, that is, the CDIO Syllabus is more comprehensive and has a more explicit organization that is based on the tasks of engineering. Consequently, an engineering education program designed to meet the student learning outcomes set forth in the Syllabus can easily meet its respective national standards. Alignment with the objectives of the Bologna Declaration is discussed in Chapter Eleven. The CDIO Syllabus outcomes and the 12 CDIO Standards are stretch goals that even the best programs around the world must work diligently to meet. National standards present the *rules* of what to do. By contrast, the Standards and Syllabus form a best-practice framework that serves as a *playbook*—the approaches, resources, and community that allow a program to achieve its goals.

Attracting and motivating students who are "ready to engineer"

One of the important requirements of the CDIO Initiative is to make engineering more interesting, and therefore increase student motivation and retention. In much of the developed world, and in the developing world as well, there is great concern that more scientists and technologists will be needed in the future, and that current supply is insufficient. We believe that we have incorporated several features that will attract and motivate students.

Many students are attracted to engineering by the belief that engineers build things and are disappointed by the first years of traditional engineering education, when they are taught theory. By placing early and repeated design-implement experiences in the curriculum, we have appealed to this desire to build and create. Many students complain that engineering education "beats them down" through a demanding schedule of theory-rich education with little reward. By using active and experiential learning techniques and projects, we offer students a chance to develop a sense of empowerment and self-efficacy critical to their perception of self-worth. Projects also provide outlets for creativity and leadership, with visible signs of accomplishment. These factors are captured in the reaction of several students who have graduated from our programs. Their experiences are framed in Box 2.6.

Box 2.6. Student Views of the Benefit of a CDIO Program

The single reason I picked KTH over another school was the promise of building an aircraft at the end of the program—something the other schools didn't offer. A course where you get to design and build and fly is a great opportunity to try your own wings, to see how much you've actually learned, and to own the whole process. It is much more rewarding to solve your own problem, instead of the professor's problem sets. To practice skills and technical knowledge in a project makes you feel more ready for the real job of engineering.

The motivation for a program of study has to be readily visible. Therefore, it is a good thing that the first-year course has a motivating design experience. At the same time, different professors introduce their respective fields of work and what you need to get there yourself. I, for one, needed to have those ideas fresh in my mind while struggling with the calculus course.

The list of skills in the CDIO Syllabus may be long, but students agree that it is what you need in the workplace—communication, ethics, and societal context. However, engineering students don't want to take separate courses on each of these knowledge and skill areas. They appreciate an integrated curriculum, where skills are interlaced with engineering knowledge and practice. This is, in fact, how these skills are encountered in real life. All these reasons combine to make CDIO an engineering education that makes you ready to engineer.

– H. Grankvist, Former Student, KTH - Royal Institute of Technology

One of the major benefits of participating in a CDIO program is that it allows you to develop skills such as *engineering reasoning* and *problem solving*. Our profession demands that you have the ability to identify and formulate problems, as well as formulate solutions and recommendations. These are essential skills that a CDIO program emphasizes. I find that the skills taught are very important, both for me personally and also for my future employers. The skills of engineering reasoning and problem solving also help bridge the gap between university study and work life, making the transition easier and quicker.

A CDIO program creates a supportive environment for today's engineers as we prepare to be a part of a profession where *teamwork* and *communication skills* are essential. In a way, a CDIO program assures a certain level of development in these skills. Consequently, all students, not only the students who are most active in extra-curricular activities, are able to develop these skills during their university years. I believe that we are personally responsible for our own development. By taking part in a CDIO program, we learn the importance of this at an early stage.

The development of skills has not reduced the importance of underlying sciences (mathematics, physics, chemistry, biology, etc) and core engineering fundamental sciences. After all, it is the combination of engineering reasoning, problem solving, teamwork, and communication skills, taken together with knowledge of core engineering fundamentals that makes us attractive as future engineers.

– A. Wibring, Former Student, Chalmers University of Technology

In my view, the ideal engineering program is well described by the CDIO Syllabus. The emphasis is on technical knowledge and practical methods, which are taught in the context of the real-world requirements of the engineering profession. Teamwork, written communication, and professional ethics, as well as an understanding of the external (*e.g.*, financial, political, environmental) factors that affect today's engineers are important features of the curriculum.

During my education, I was able to develop many of the skills a CDIO program is intended to address. Early in the program, coursework emphasized knowledge of the engineering sciences and its application in problem solving. Later courses included more of the "new" elements of the curriculum, such as working in project teams and delivering presentations. In general, these assignments were a valuable part of my engineering studies and have paid dividends since graduation.

Beyond a solid technical foundation, I have drawn on other skills in my current work toward a Master's degree in Aeronautics and Astronautics. My thesis work is in space vehicle guidance, a rather narrow field, and I have relied on a systems-level perspective to ensure that my work will be relevant in the design of future space systems. The ability to understand the impact of subsystem design choices on the overall system was one of my notable takeaways from the program. The ability to collaborate with others and to convey the essence of a problem is another example of a skill I trace directly to my experience in a CDIO program.

– P. Springmann, Former Student, Massachusetts Institute of Technology

Another factor in attracting and motivating students is to show that the education leads to higher quality employment. In fact, by responding to the input of industry stakeholders who hire our students, we should be preparing students who are "ready to engineer"—more readily hired, have more successful careers, and more impact in their profession. Preliminary indications are that firms familiar with the CDIO Initiative are eager to hire graduates of these programs as evidenced by the comments of Billy Fredriksson, former Chief Technology Officer of SAAB, presented in Box 2.7. If we make engineering education more interesting, empowering, and rewarding, and simultaneously increase the learning of both fundamentals and skills, the demand for this education will increase and the needs of society for a technological workforce will be met.

Box 2.7. CDIO Engineers in Industry

Industry would prefer to hire engineers from CDIO programs because they have received excellent training in how to apply their basic theoretical knowledge to the development of practical product or process-related projects. During their studies, CDIO engineers get a good introduction to the real practice of engineering. They have learned both the technical skills and also personal and interpersonal skills; and the importance of holistic approaches and systems integration in designing and building products. This means that the CDIO engineers will probably be able to apply their knowledge more quickly when starting work in industry. They can more easily and quickly work productively in engineering teams.

There are several reasons why engineers graduating from a CDIO program will likely have more options and be more successful in pursuing their careers. I would expect CDIO engineers to start their industrial careers more rapidly, either as a disciplinary specialist or as a project engineer. As disciplinary specialists, they know the importance of taking into account requirements from related areas when integrating results into the product or system. As project engineers or project leaders, they are more prepared for, and understand the importance of, teamwork and other personal and interpersonal skills. They are able to look after and secure the integrated result and performance of the final product, and they recognize the importance of timing to the project.

Thus, graduates from CDIO programs will be more attractive to industry and more likely to succeed both personally, and in their responsibility to build systems of value to society.

– B. Fredriksson, Saab

Meeting the requirements

In this section, we have demonstrated that the CDIO Initiative meets the remaining four requirements for successful reform of engineering education:

- It is comprehensive in its reform of all important student learning experiences.
- It is a consortium of programs and departments that collaborate in parallel development and share resources.
- It retasks existing resources during ongoing operation.
- It is designed to attract, retain, and graduate qualified students for the profession.

SUMMARY

This chapter has given an overview of the CDIO Initiative, its goals, context, vision and pedagogical foundation, and in doing so, the chapter has demonstrated that it meets the eight requirements for successful reform of engineering education. We have also outlined the answers provided to the two central questions that any approach to improving engineering education must address:

- *What is the full set of knowledge, skills, and attitudes that engineering students should possess as they leave the university, and at what level of proficiency?*
- *How can we do better at ensuring that students learn these skills?*

The first question is addressed by the CDIO Syllabus and the process for reaching stakeholder consensus on the level of proficiency that students should attain in a given program. The Syllabus is discussed in more detail in Chapter Three.

The second question is addressed in Chapters Four through Seven, which discuss curriculum design, design-implement experiences, teaching and learning, and student assessment, using the CDIO Standards as the organizing principle. Chapters Eight and Nine discuss program evaluation and the change process necessary to enhance faculty competence and to lead a successful adaptation and implementation of the CDIO approach to a university program.

Our treatment of engineering education concludes by looking backward and forward. Chapter Ten sets the effort in the historical context of engineering education reform, and Chapter Eleven looks ahead to the challenges of engineering education in the next decades.

DISCUSSION QUESTIONS

1. In what ways are you improving engineering education in your own programs?
2. How can the CDIO approach to engineering education be applied to your reform initiatives?
3. Which barriers to educational reform are common to programs around the world? Which may be unique to your program?
4. How do your educational initiatives compare with the CDIO approach and other reform efforts?

References

[1] Von Kármán, T., In A. L. Mackay, *Dictionary of Scientific Quotations,* London, 1994.

[2] The Royal Charter, The Institution of Civil Engineers, London, 1828. Available at http://www.ice.org.uk

[3] Finiston, M., *Engineering Our Future: Report of the Committee of Inquiry into the Engineering Profession*, HMSO CMND 7794, London, 1980.

[4] Gordon, B. M., "What is an Engineer?", Invited Keynote Presentation, Annual Conference of the European Society for Engineering Education (SEFI), University of Erlangen-Nürnberg, 1984.

[5] The Boeing Company, *Desired Attributes of an Engineer,* 1996. Available at http://www.boeing.com/companyoffices/pwu/attributes/attributes.html

[6] The National Academy of Engineering. Center for the Advancement of Scholarship in Engineering Education (CASEE). Available at http://www.nae.edu

[7] Wiggins, G., and McTighe, J., *Understanding by Design,* Association for Supervision and Curriculum Development, Alexandria, Virginia, 1998.
[8] Marton, F., and Säljö., R., Approaches to Learning, in Marton, F., Hounsell, D., and Entwistle, N. J., Eds., *The Experience of Learning*. Edinburgh: Scottish Academic Press, 1984.
[9] Gibbs, G., *Improving the Quality of Student Learning*, TES, Bristol, England, 1992.
[10] Rhem, J., *National Teaching and Learning Forum*, Vol. 5, No. 1, 1995.
[11] Biggs, J., *Teaching for Quality Learning At University*, 2nd ed., The Society for Research into Higher Education and Open University Press, Berkshire, England, 2003.
[12] Jarvis, P., Holford, J., and Griffin, C., *The Theory and Practice of Learning*, Kogan Page, London, 1998.
[13] Brainerd, C. J., Piaget, J., Learning, Research, and American Education, in Zimmerman, B. J., and Schunk, D. H., *Educational Psychology: A Century of Contributions*, Lawrence Erlbaum Associates, London, 2003.
[14] Kolb, D. A., *Experiential Learning*, Prentice-Hall, Upper Saddle River, New Jersey, 1984.
[15] Accreditation Board of Engineering and Technology, *Criteria for Accrediting Engineering Programs: Effective for Evaluations During the 2000-2001 Accreditation Cycle*, 2000. Available at http://www.abet.org
[16] Engineering Council, *UK Standards for Professional Engineering Competence: the Accreditation of Higher Education Programs*, 2004. Available at http://www.iee.org/professionalregistration/ukspec.cfm
[17] The Bologna Declaration. Available at http://www.crue.org/eurec/bolognaexplanation.htm
[18] The EUR-ACE Project. Available at http://www.eurace.org

CHAPTER THREE THE CDIO SYLLABUS: LEARNING OUTCOMES FOR ENGINEERING EDUCATION

WITH P. J. ARMSTRONG

INTRODUCTION

We will now develop a comprehensive approach to answering the first of the two questions, central to the reform of engineering education, posed in Chapter Two.

> *What is the full set of knowledge, skills, and attitudes that engineering students should possess as they leave the university, and at what level of proficiency?*

Said another way, what are the desired learning outcomes for engineering education?

This question highlights the tension between two apparently conflicting needs—one that has its origin in the recent history of engineering education, as described in Chapter Two; the other, the ever-increasing body of technical knowledge that graduating students must command. It is our responsibility as university educators to introduce them to this broad body of disciplinary knowledge. On the other hand, engineers must possess a wide array of personal and interpersonal skills; and product, process and system building skills that will allow them to function in real engineering teams and to produce tangible benefits to society.

The CDIO Initiative has developed an educational approach that attempts to resolve this tension and to address the complete needs of our students. This approach entails first developing a comprehensive understanding of the knowledge, skills, and attitudes needed by the contemporary engineer, that is, the desired learning outcomes. The development of this understanding is the subject of this chapter. The curricular, pedagogical, and assessment strategies to facilitate meeting these learning outcomes are addressed in the next six chapters of this book.

This chapter describes the development and content of the CDIO Syllabus, a codification of contemporary engineering knowledge, skills, and attitudes that constitute the foundation for the reform of university engineering education programs. Engineers might view the Syllabus as a *requirements document* for engineering education. For education specialists, it will be viewed as a comprehensive statement of learning outcomes. Both are equally valid interpretations.

It is our aim that we move toward a resolution of the tension in contemporary engineering education by providing a complete enumeration of the knowledge and skills that graduating students should possess. This enumeration should be sufficiently general to allow it to be applied to all branches of engineering. It should be sufficiently detailed to be useful in curriculum planning and learning assessment. The first half of the chapter describes the development of the Syllabus, addressing the first part of the central question, "*What is the full set of knowledge, skills, and attitudes that engineering students should possess as they leave the university?*"

Traditionally, the second part of the central question—*at what level of proficiency?*—is decided internally by university faculty, by consensus, or by the choice of individual instructors. We advocate an approach that includes stakeholders from among students, faculty, university staff, alumni, and industry representatives in consensus to set the expected level of proficiency for each learning outcome. There is nothing that limits the two essential steps in this process—complete enumeration of outcomes and stakeholder engagement in setting expected levels of proficiency—from applying to any educational endeavor. The chapter also presents a generalization of this process to broad areas of education.

CHAPTER OBJECTIVES

This chapter is designed so that you can

- explain how the content of the CDIO Syllabus is derived from engineering practice
- describe the content and structure of the Syllabus
- explain the rationale for specifying learning outcomes in personal and interpersonal skills; and product, process and system building skills, as well as in technical disciplines
- describe how to engage stakeholders within and outside the university in the development of detailed learning outcomes
- outline a process for developing learning outcomes for engineering education that can be generalized to all disciplines

THE KNOWLEDGE AND SKILLS OF ENGINEERING

The required knowledge and skills of engineering are best defined through the examination of the practice of engineering. In fact, from its conception as a profession early in the 19th century until the middle of the 20th century, engineering education was based on engineering practice. As explained in Chapter

Two, the last half-century of engineering education saw its development move from a practice base to an engineering-science base. We now are observing a renewed interest in developing a third approach that merges the best of the engineering-science and practice viewpoints. This third approach requires a re-examination of the needs of modern engineering practice.

Required engineering knowledge and skills

As early as the 1940s, attempts were made to codify the nontraditional skills an engineer must possess. One such attempt, the *Unwritten Laws of Engineering* [1], called for the development of such skills as oral and written communication, planning, and the ability to work successfully in organizations. In addition, the *Unwritten Laws* emphasized the importance of personal attributes, such as propensity toward action, integrity, and self-reliance. In many ways, this list of skills remains as valid for today's engineers as it was over a half century ago.

With the advent of the modern engineering-science approach in the 1950s, the education of engineers became more disassociated from the practice of engineering. Engineering science became the dominant culture of engineering schools, where fewer faculty members worked as engineers prior to teaching. By the 1980s, engineering educators and industrialists began to react to this widening gulf between engineering education and practice. For example, in the essay entitled *What is an Engineer?* [2], Gordon clearly enumerates the knowledge and skills required for contemporary engineering practice. (See Box 2.1 in Chapter Two.)

The past decade has seen a concerted effort to close the gap between engineering education and practice. Major engineering companies (*e.g.*, Boeing) published lists of desired attributes [3], and leaders of industry urged a new look at the issues related to the qualifications of engineers [4]. One might ask if these lists are particular to the United States, a specific field of engineering, or the needs of a decade? It is interesting that ten years after these two references appeared, the World Chemical Engineering Council produced the 2004 list of the evident shortcomings of engineering graduates with respect to important skills of engineering graduates [5], as shown in Table 3.1. Comparison of

TABLE 3.1. EVIDENT SHORTCOMINGS OF GRADUATING ENGINEERS WITH RESPECT TO SKILLS AND ABILITIES

Most important abilities with respect to EMPLOYMENT	Greatest deficits in abilities with respect to EDUCATION
Work effectively as a team	Business approach
Analyze information	Management skills
Communicate effectively	Project management methods
Gather information	Methods for quality assurance
Self-learning	Ability to communicate effectively
	Knowledge of marketing principles
	Sense of ethical and professional responsibilities

this list with those produced by Boeing (see Box 2.2 in Chapter Two), and the Accreditation Board for Engineering and Technology—the ABET EC2000 criteria [6]—as well as other sources spanning fifty years yields a remarkably consistent image of the desired attributes of young engineers. The required knowledge, skills, and attitudes that companies desire in their engineers consistently include an understanding of engineering fundamentals, design, and manufacturing; the context of engineering practice; and the ability to think critically and creatively, to communicate, and to work on teams.

Based on this consistency, industrial leaders in the United States successfully lobbied government agencies to fund science and engineering education reform, persuaded professional societies to revise accreditation standards, and created joint working groups to facilitate the exchange of viewpoints. Other industrialized countries around the world initiated similar educational reforms. Despite good intentions, most of these initiatives did not have the fundamental impact on education originally desired.

Importance of rationale and levels of detail

Two key reasons account for the lack of convergence between engineering education and engineering practice: an absence of rationale and an absence of detail. Previous lists of skills were derived requirements that failed to make convincing statements of the rationale for why these were the desired attributes of an engineer. As explained in Chapter Two, the CDIO approach reformulates the underlying need to make the rationale more explicit. Therefore, the starting point of our effort was a restatement of the underlying *need* for engineering education. We believe that every graduating engineer should be able to:

> *Conceive-Design-Implement-Operate complex value-added engineering products, processes, and systems in a modern, team-based environment.*

The rationale is essentially a restatement of the fact that it is the job of engineers to be able to engineer. If the conceive-design-implement-operate premise is accepted as the context of engineering education, it is possible to derive more detailed goals and learning outcomes for engineering education that are understandable within this rationale.

The second limitation is the fact that other existing lists of skills lack sufficient detail to be widely understood or implemented. The CDIO Syllabus was developed to address this limitation by creating a clear, complete, and consistent set of goals for engineering education, in sufficient detail that they can be understood and implemented by engineering faculty. This set of detailed goals forms the basis for rational design of the curriculum and a comprehensive system of assessment.

The formulation of the functions of an engineer, from which the Syllabus is derived, does not in any way diminish the role of engineering science or engineering research. On the contrary, engineering science is the appropriate basis for engineering education, and engineering research is the process of adding new knowledge to that base. Although most collaborators in the CDIO Initiative are engineering scientists and researchers, our programs educate students, the vast majority of whom will go on to become professional engineers. This is true even at research-intensive universities, such as MIT in the United States and KTH in Sweden. Whether students become practicing engineers or engineering researchers, setting their educational experiences in the context of the conception, design, implementation, and operation of systems and products strengthens their backgrounds.

THE CDIO SYLLABUS

The CDIO Syllabus is a list of knowledge, skills, and attitudes rationalized against the norms of contemporary engineering practice, comprehensive of all known skills lists, and reviewed by experts in many fields. The principal value of the Syllabus is that it can be applied across a variety of programs and can serve as a model for all programs to derive specific learning outcomes. CDIO Standard 2 emphasizes the importance of the Syllabus in engineering education reform.

> STANDARD 2 – **LEARNING OUTCOMES**
> **Specific, detailed learning outcomes for personal and interpersonal skills, and product, process and system building skills, as well as disciplinary knowledge, consistent with program goals and validated by program stakeholders**

The knowledge, skills, and attitudes intended as a result of engineering education, that is, the learning outcomes are codified in the CDIO Syllabus. These learning outcomes detail what students should know and be able to do at the conclusion of their engineering programs. In addition to learning outcomes for technical disciplinary knowledge (Section 1), the Syllabus specifies learning outcomes as personal, and interpersonal; and product, process and system building. Personal learning outcomes (Section 2) focus on individual students' cognitive and affective development, which include engineering reasoning and problem solving, experimentation and knowledge discovery, system thinking, creative thinking, critical thinking, and professional ethics. Interpersonal learning outcomes (Section 3) focus on individual and group interactions such as teamwork, leadership, and communication. Product, process, and system building skills (Section 4) focus on conceiving, designing, implementing, and operating products, processes, and systems in enterprise, business, and societal contexts.

Learning outcomes are reviewed and validated by key stakeholders—groups who share an interest in the graduates of engineering programs—for consistency with program goals and relevance to engineering practice.

In addition, stakeholders help to determine the expected levels of proficiency, or standards of achievement, for each learning outcome. The process of validating the Syllabus with stakeholders is discussed later in this chapter.

Development and integration of the CDIO Syllabus

The content and structure of the Syllabus is the focus of this chapter. This content and structure were motivated, in part, by an understanding of how it will be used. The Syllabus, customized with results of stakeholder surveys, lays the foundation for curriculum planning and integration, teaching and learning practice, and outcomes-based assessment. Figure 3.1 illustrates the development of the CDIO Syllabus from needs to goals, its customization to program goals, and the integration of program goals into the curriculum. More details on this process are given in later chapters. The process of integrating the Syllabus into a program's curriculum is the subject of Chapter Four. Approaches to teaching and learning the content of the Syllabus are described in Chapter Six. Student assessment of learning outcomes is the focus of Chapter Seven.

Content and structure of the CDIO Syllabus

Three goals motivated the choice of content and structure of the Syllabus. These goals were to

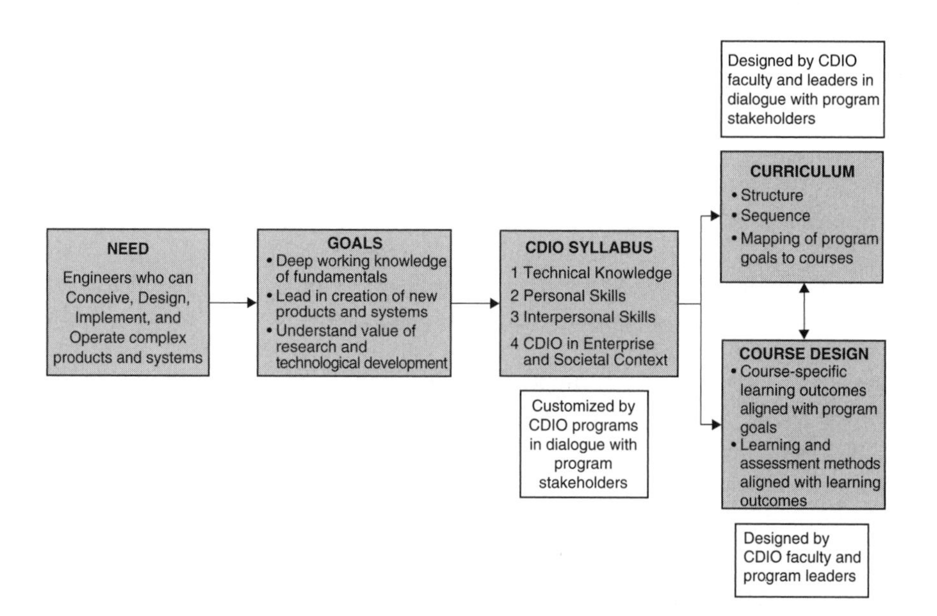

FIGURE 3.1. DEVELOPMENT AND INTEGRATION OF THE CDIO SYLLABUS

- Create a structure whose rationale is clearly visible
- Derive a comprehensive high-level set of goals that correlated with other respected sources
- Develop a clear, complete, and consistent set of topics to facilitate implementation and assessment

The point of departure for the derivation of the content of the Syllabus is the simple statement that engineers engineer, that is, they build products, processes, and systems for the betterment of humanity. In order to enter the contemporary profession of engineering, students must be able to perform the essential functions of an engineer. As described previously, graduating engineers should be able to conceive-design-implement-operate complex value-added engineering products, processes, and systems in a modern, team-based environment. Stated another way, graduating engineers should appreciate the engineering process; be able to contribute to the development of engineering products, processes, and systems; and do so while working in engineering organizations. Implicit is the additional expectation that, as university graduates and young adults, engineering graduates should be mature and thoughtful individuals.

These high-level expectations map directly to the first-level, or X-level, organization of the Syllabus, as illustrated in Figure 3.2. The mapping of the first-level Syllabus items to the four expectations illustrates that a mature individual interested in technical endeavors possesses a set of *Personal and Professional Skills and Attributes*, central to the practice. In order to develop complex, value-added engineering systems, students must have mastered the fundamentals of the appropriate *Technical Knowledge and Reasoning*. In order to work in modern, team-based environments, students must have developed the *Interpersonal Skills* of teamwork and communication. Finally, in order to create and operate products, processes, and systems, students must understand something of *Conceiving, Designing, Implementing, and Operating Systems in the Enterprise and Societal Context*.

The second level of detailed content, or X.X, of *1 Technical Knowledge and Reasoning* is shown in Figure 3.3. Modern engineering relies on *Knowledge of*

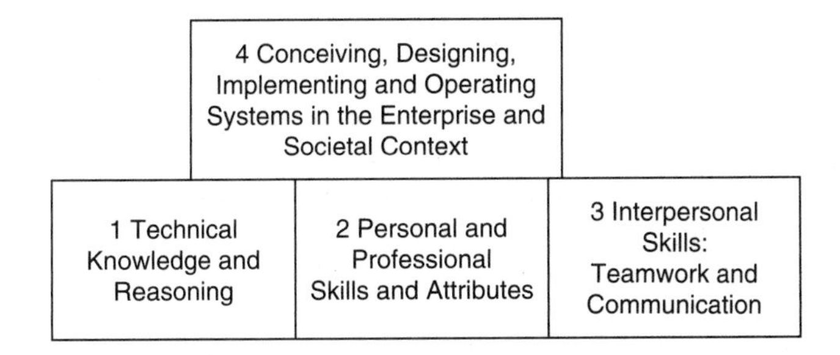

FIGURE 3.2. **THE CDIO SYLLABUS AT THE FIRST LEVEL OF DETAIL**

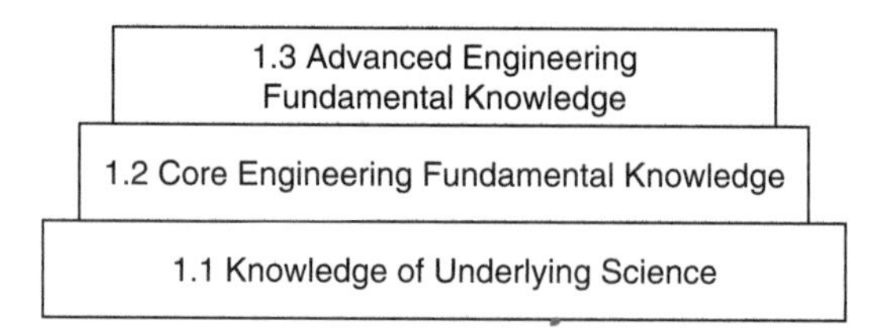

FIGURE 3.3. THE **CDIO** SYLLABUS: TECHNICAL KNOWLEDGE AND REASONING

Underlying Sciences (1.1). A body of *Core Engineering Fundamental Knowledge* (1.2) builds on that science core, and a set of *Advanced Engineering Fundamentals* (1.3) moves students towards the skills necessary to begin a professional career. This is the disciplinary curriculum that engineering school faculty usually debate and define. This section of the Syllabus is, in fact, a placeholder for the more detailed description of the disciplinary fundamentals necessary for engineering education. The details of Section 1 vary widely in content from field to field. The placement of *Technical Knowledge and Reasoning* at the beginning of the Syllabus is a reminder that the development of a deep working knowledge of technical fundamentals is, and should be, the primary objective of engineering education. The remainder of the Syllabus addresses the more generic knowledge, skills, and attitudes that all engineering graduates should possess.

Engineers of all types use approximately the same personal and interpersonal skills, and follow approximately the same generalized processes. We have tried in the remaining three parts of the Syllabus to be inclusive of all the knowledge, skills, and attitudes that engineering graduates might require. In addition, we have attempted to use terminology recognizable to all professions. Local usage in different engineering fields will require some translation and interpretation.

The second-level content of *2 Personal and Professional Skills and Attributes* is shown with *3 Interpersonal Skills* in Figure 3.4. The innermost circle highlights the three modes of thought most practiced by engineers: *Engineering Reasoning and Problem Solving* (2.1), *Experimentation and Knowledge Discovery* (2.2), and *System Thinking* (2.3). These might also be called engineering thinking, scientific thinking, and system thinking. Each mode of thinking is further detailed into formulation of issues, the process of thinking, and resolution of issues. *Professional Skills and Attitudes* (2.5), other than the three modes of thought, include professional integrity, professional behavior, and the skills and attitudes necessary to plan for careers and lifelong learning in the world of engineering. *Personal Skills and Attitudes* (2.4) include general character traits of initiative and perseverance, creative and critical thinking, self-knowledge, curiosity and lifelong learning, and time management.

Interpersonal Skills are a distinct subset of personal skills that divide into three overlapping subsets: *Multidisciplinary Teamwork* (3.1), *Communications*

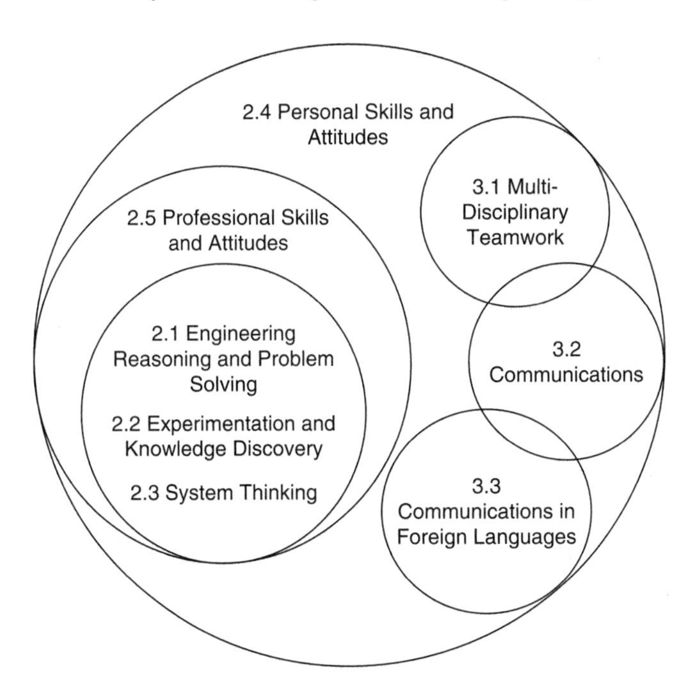

FIGURE 3.4. THE CDIO SYLLABUS: PERSONAL, PROFESSIONAL, AND INTERPERSONAL SKILLS

(3.2) and *Communications in Foreign Languages* (3.3). Teamwork is comprised of forming, operating, growing, and leading technical teams. Communications includes the skills necessary to devise a communications strategy and structure, and to use four common modes of communication—written, oral, graphic, and electronic. *Communications in Foreign Languages* includes traditional skills associated with foreign language learning, and applications made specifically for technical communications.

Figure 3.5 shows an overview of *4 Conceiving, Designing, Implementing, and Operating Systems in the Enterprise and Societal Context*. It illustrates the way the development of a product, process, or system moves through four phases: *Conceiving and Engineering Systems* (4.3), *Designing* (4.4), *Implementing* (4.5), and *Operating* (4.6). The terms are chosen to be descriptive of hardware, software, systems, and process industries. *Conceiving and Engineering Systems* takes the process from market or opportunity identification through high-level conceptual design and includes development and project management. *Designing* includes aspects of the design process, as well as disciplinary, multidisciplinary, and multi-objective design. *Implementing* includes hardware and software processes; test and verification; as well as design and management of the implementation process. *Operating* covers a wide range of issues from designing and managing operations, through supporting the product, process, and system lifecycle and improvement, to the end-of-lifecycle planning.

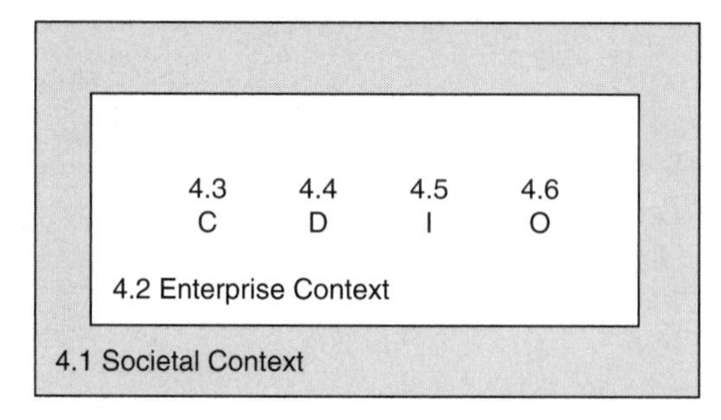

FIGURE 3.5. **THE CDIO SYLLABUS: CONCEIVING, DESIGNING, IMPLEMENTING, AND OPERATING**

Products, processes, and systems are created and operated within an *Enterprise and Business Context* (4.2) that engineers must understand to operate effectively. The skills necessary to do this include recognizing the culture and strategy of an enterprise and understanding how to act in an entrepreneurial way within an enterprise of any type or size. Likewise, enterprises exist within a larger *External and Societal Context* (4.1). Knowledge and skills in this area include acknowledgment of the relationship between society and engineering, and an understanding of the broader historical, cultural, and global context.

In summary, the Syllabus is organized at the first two levels in rational manner. The first level, or X level, reflects the functions of an engineer, who is a well-developed individual, involved in a process that is embedded in an organization, with the intent of building products, processes, and systems. The second level of detailed content, or X.X level, reflects contemporary practice and scholarship of the engineering profession.

The Syllabus is defined to third and fourth levels of detail, respectively, the X.X.X level and the X.X.X.X level. These fine-grain details are necessary to transition from high-level goals to teachable and assessable learning outcomes. Although it could seem overwhelming at first, the detailed Syllabus has many benefits for engineering faculty who may not be experts in some of the Syllabus topics. The details provide insight into content and learning outcomes, the integration of these skills into a curriculum, and the planning of teaching and assessment. Table 3.2 is a condensed version at the third level of detail. Appendix A is the complete CDIO Syllabus at the fourth level of detail.

Validation of the CDIO Syllabus

The process used to arrive at the detailed content of the Syllabus blended elements of user-need studies in product development with other techniques used in scholarly research. It included focus-group discussions, document research, surveys, workshops, and peer review. Focus groups were conducted with faculty, students, industry leaders, and senior engineering academics from a variety of universities. To ensure applicability to all engineering fields,

TABLE 3.2. CONDENSED CDIO SYLLABUS AT THE THIRD LEVEL OF DETAIL

1 TECHNICAL KNOWLEDGE AND REASONING

1.1 KNOWLEDGE OF UNDERLYING SCIENCE
1.2 CORE ENGINEERING FUNDAMENTAL KNOWLEDGE
1.3 ADVANCED ENGINEERING FUNDAMENTAL KNOWLEDGE

2 PERSONAL AND PROFESSIONAL SKILLS AND ATTRIBUTES

2.1 ENGINEERING REASONING AND PROBLEM SOLVING
- 2.1.1 Problem Identification and Framing
- 2.1.2 Modeling
- 2.1.3 Estimation and Qualitative Analysis
- 2.1.4 Analysis With Uncertainty
- 2.1.5 Closing the Problem

2.2 EXPERIMENTATION AND KNOWLEDGE DISCOVERY
- 2.2.1 Principles of Research and Inquiry
- 2.2.2 Experimental Inquiry
- 2.2.3 Survey of Print and Electronic Literature
- 2.2.4 Hypothesis Test, and Defense

2.3 SYSTEM THINKING
- 2.3.1 Thinking Holistically
- 2.3.2 Emergence and Interactions in Systems
- 2.3.3 Prioritization and Focus
- 2.3.4 Trade-offs and Balance

2.4 PERSONAL SKILLS AND ATTITUDES
- 2.4.1 Initiative and Willingness to Take Risks
- 2.4.2 Perseverance, and Flexibility
- 2.4.3 Creative Thinking
- 2.4.4 Critical Thinking
- 2.4.5 Personal Inventory
- 2.4.6 Curiosity and Lifelong Learning
- 2.4.7 Time and Resource Management

2.5 PROFESSIONAL SKILLS AND ATTITUDES
- 2.5.1 Professional Ethics, Integrity, Responsibility and Accountability
- 2.5.2 Professional Behavior
- 2.5.3 Proactively Planning for One's Career
- 2.5.4 Stay Current on World of Engineer

3 INTERPERSONAL SKILLS: TEAMWORK AND COMMUNICATION

3.1 MULTI-DISCIPLINARY TEAMWORK
- 3.1.1 Form Effective Teams
- 3.1.2 Team Operation
- 3.1.3 Team Growth and Evolution
- 3.1.4 Leadership
- 3.1.5 Technical Teaming

3.2 COMMUNICATIONS
- 3.2.1 Communications Strategy
- 3.2.2 Communications Structure
- 3.2.3 Written Communication
- 3.2.4 Electronic/Multimedia Communication
- 3.2.5 Graphical Communication
- 3.2.6 Oral Presentation and Interpersonal Communications

3.3 COMMUNICATIONS IN FOREIGN LANGUAGES
- 3.3.1 English
- 3.3.2 Languages of Regional Industrial Nations
- 3.3.3 Other Languages

4 CONCEIVING, DESIGNING, IMPLEMENTING, AND OPERATING SYSTEMS IN THE ENTERPRISE AND SOCIETAL CONTEXT

4.1 EXTERNAL AND SOCIETAL CONTEXT
- 4.1.1 Roles and Responsibilities of Engineers
- 4.1.2 Understand the Impact of Engineering
- 4.1.3 Understand How Engineering is Regulated
- 4.1.4 Knowledge of Historical and Cultural Context
- 4.1.5 Knowledge of Contemporary Issues and Values
- 4.1.6 Developing a Global Perspective

4.2 ENTERPRISE AND BUSINESS CONTEXT
- 4.2.1 Appreciating Different Enterprise Cultures
- 4.2.2 Enterprise Strategy, Goals, and Planning
- 4.2.3 Technical Entrepreneurship
- 4.2.4 Working Successfully in Organizations

(*Continued*)

TABLE 3.2. CONDENSED CDIO SYLLABUS AT THE THIRD LEVEL OF DETAIL—CONT'D

4.3 CONCEIVING AND ENGINEERING SYSTEMS	4.5 IMPLEMENTING
4.3.1 Setting System Goals and Requirements	4.5.1 Designing and Modeling of the Implementation Process
4.3.2 Defining Function, Concept and Architecture	4.5.2 Hardware Manufacturing Process
4.3.3 Modeling of System and Ensuring Goals Can Be Met	4.5.3 Software Implementing Process
4.3.4 Project Management	4.5.4 Hardware Software Integration
4.4 DESIGNING	4.5.5 Test, Verification, Validation, and Certification
4.4.1 The Design Process	4.5.6 Managing Implementation
4.4.2 The Design Process Phasing and Approaches	4.6 OPERATING
4.4.3 Utilization of Knowledge in Design	4.6.1 Modeling, Designing and Optimizing Operations
4.4.4 Disciplinary Design	4.6.2 Training and Operations
4.4.5 Multidisciplinary Design	4.6.3 Supporting the System Lifecycle
4.4.6 Multi-Objective Design (DFX)	4.6.4 System Improvement and Evolution
	4.6.5 Disposal and Life-End Issues
	4.6.6 Operations Management

individuals with varied engineering backgrounds were included. The groups were presented with the question, *"What, in detail, is the set of knowledge, skills, and attitudes that a graduating engineer should possess?"*

Results of the focus groups were combined with topics extracted from four principal, comprehensive source documents into a preliminary draft. The source documents represented the views of industry, government, and higher education on the expectations for university graduates. They included the aforementioned ABET EC2000 criteria [6] and Boeing's *Desired Attributes of an Engineer* [3], as well as two documents from the Massachusetts Institute of Technology [7]-[8].

To obtain stakeholder feedback that would validate the Syllabus topics, a survey was conducted among four constituencies: faculty, senior industry leaders, younger alumni (average age of 25) and older alumni (average age of 35). The qualitative comments from this survey were incorporated, improving the Syllabus' organization, clarity and coverage. Subsequently, several domain experts reviewed each second-level, or X.X-level, Syllabus topic. Combining the results of the expert reviews and additional references, a final version of the Syllabus topics was drafted.

To ensure comprehensiveness and to facilitate comparisons, the Syllabus content was correlated with the four comprehensive source documents referred to earlier. For example, the Syllabus topics at the second level were correlated with ABET's EC2000 Criteria 3a to 3k. (See Table 3.3) EC2000 states that accredited engineering programs must ensure that its graduates have developed 11 specific attributes [6]. The correlation of the Syllabus with these 11 attributes is strong. In fact, the Syllabus is more comprehensive. For example, EC2000 does not explicitly address *System Thinking* (2.3), and lists only *an ability to engage in lifelong learning* (3i) from among the many desirable *Personal Skills and Attitudes* (2.4) of the Syllabus. Likewise, EC2000 lists only *an understanding*

TABLE 3.3. **THE CDIO SYLLABUS CORRELATED WITH ABET EC2000 CRITERION 3**

a. An ability to apply knowledge of mathematics, science, and engineering.
b. An ability to design and conduct experiments, as well as to analyze and interpret data.
c. An ability to design a system, component, or process to meet desired needs.
d. An ability to function on multi-disciplinary teams.
e. An ability to identify, formulate, and solve engineering problems.
f. An understanding of professional and ethical responsibility.
g. An ability to communicate effectively.
h. The broad education necessary to understand the impact of engineering solutions in a global and societal context.
i. A recognition of the need for, and an ability to engage in, life-long learning.
j. A knowledge of contemporary issues
k. An ability to use the techniques, skills, and modern engineering tools necessary for engineering practice

CDIO SYLLABUS	ABET EC2000 CRITERION 3										
	a	b	c	d	e	f	g	h	i	j	k
1.1 Knowledge of Underlying Science	■										
1.2 Core Engineering Fundamentals	■										
1.3 Advanced Engineering Fundamental Knowledge	□										■
2.1 Engineering Reasoning and Problem Solving											□
2.2 Experimentation and Knowledge Discovery		■									
2.3 System Thinking			□								
2.4 Personal Skills and Attitudes									■		
2.5 Professional Skills and Attitudes						■					□
3.1 Multi-disciplinary Teamwork				■							
3.2 Communications							■				
3.3 Communications in Foreign Languages											
4.1 External and Societal Context								■		■	
4.2 Enterprise and Business Context											
4.3 Conceiving and Engineering Systems			■								
4.4 Designing			■								
4.5 Implementing			■								
4.6 Operating			■								

■ Strong Correlation □ Good Correlation

of professional and ethical responsibility (3f) from among several important *Professional Skills and Attitudes* (2.5).

The ABET document comes closer than other source documents to capturing the full involvement in a product lifecycle by specifying item (3c) *the ability to design a system, component, or process to meet desired needs*. Designing a system to meet desired needs hints at the spirit of *Conceiving and Engineering Systems* (4.3) in the Syllabus. *Designing a component* maps to *Designing* (4.4), and *designing a process* could be construed to include *Implementing* (4.5) and *Operating* (4.6).

Comparing the CDIO Syllabus with ABET EC2000 Criterion 3, the Syllabus has two advantages, one minor and one major. The minor advantage is that the Syllabus is more rationally organized. It is explicitly derived from the functions of modern engineering. Although this organization might not provide a better understanding of how to implement change, it certainly creates a better understanding of the reasons to change. The major advantage is that the Syllabus contains more levels of detail than the ABET document. It penetrates into enough detail that general phrases such as *good communication skills* take on substantive meaning. Furthermore, it defines measurable goals that are critical to curriculum design and assessment.

Similar comparisons have been made of the CDIO Syllabus with accreditation standards in other countries. Box 3.1 compares it with standards for engineering programs in the United Kingdom [9].

Box 3.1. Comparisons of the CDIO Syllabus With UK-SPEC

The CDIO Syllabus has been compared retrospectively with accreditation criteria published in 2004 for engineering programs in the United Kingdom. The new criteria replaced requirements contained in a document known as SARTOR 3 (*Standards and Routes to Registration*, Engineering Council, 3rd Edition, 1997.). The SARTOR 3 document did not explicitly list learning outcomes for engineering education. However, it did specify curriculum content and, somewhat controversially, made the Master of Engineering (MEng) degree the minimum academic qualification for becoming a Chartered Engineer (CEng). Students graduating with the less academically demanding Bachelor of Engineering (BEng) degree would become Incorporated Engineers (IEng), unless they undertook one year of postgraduate study to match the educational attainment of MEng graduates.

SARTOR 3 was replaced initially in December 2003, with a new set of requirements called UK-SPEC, which focused on the threshold standards of competence required for registration as a Chartered or Incorporated Engineer (*UK Standard for Professional Engineering Competence: Chartered Engineer and Incorporated Engineer Standard*, Engineering Council, 2003). A subsequent document published in May 2004 dealt with the accreditation of engineering degree programs (*UK Standard for Professional Engineering Competence: The Accreditation of Higher Education Programs*, Engineering Council, 2004). The UK-SPEC document is less prescriptive than SARTOR 3, and, in line with current thinking, accreditation criteria are expressed in the form of a list of required learning outcomes. The latter are presented under the headings *General Learning Outcomes* and *Specific Learning Outcomes*.

A. General Learning Outcomes
   1. Knowledge and Understanding
   2. Intellectual Abilities
   3. Practical Skills
   4. General Transferable Skills
B. Specific Learning Outcomes
   1. Underpinning Science and Mathematics, and Associated Engineering Disciplines
   2. Engineering Analysis
   3. Design
   4. Economic, Social and Environmental Context
   5. Engineering Practice

In UK-SPEC, an initial set of learning outcomes is provided for BEng programs; then, supplementary outcomes are stipulated under most headings for MEng programs. The learning outcomes are more detailed than the ABET EC2000 requirements. As an illustration, 40 separate learning outcomes are listed under *Specific Learning Outcomes* for an MEng program. However, in a number of cases, outcomes lack precision and clarity. In part, this is a consequence of the system in the United Kingdom, where degree programs are not accredited by a central body, but by individual engineering institutions, for example, the Institution of Mechanical Engineers and the Institution of Electrical Engineers. There are more than 30 licensed institutions. Following the publication of SARTOR 3, many produced interpretive documents in order to customize the accreditation criteria for their particular disciplines. The expectation is that this will be repeated in the case of UK-SPEC.

The outcomes listed as *General Learning Outcomes* in UK-SPEC are, in most cases, reproduced and expanded in the *Specific Learning Outcomes*. The exception is *General Transferable Skills* where reference is made to the *higher-level key skills*, defined by the Qualifications and Curriculum Authority. This body provides extensive guidance on the development of key skills in six areas: application of number, communication, information, and communication technology, improving one's own learning and performance, problem solving, and working with others. Levels of attainment are set for all areas of the education system in the UK, including university education. However, the university level key skills are generic, and do not specifically relate to engineering education.

There is a relatively weak correlation between the UK-SPEC accreditation criteria and the previously published UK-SPEC standards for professional registration. In particular, the required learning outcomes do not mirror registration standards relating to leadership, interpersonal skills, and communication in a professional setting. In part, this is a result of the decision to delegate responsibility for defining transferable skills to the Qualifications and Curriculum Authority. It may also reflect a view that professional skills and attitudes can only be acquired when graduates enter employment as practicing engineers.

As indicated in the list, UK-SPEC devotes a specific group of learning outcomes to design. Within the group there are some references to the conceptual phase that precedes design. However, the contention that engineering education should cover the complete product or system lifecycle is not fully reflected in UK-SPEC. One learning outcome states that graduates should have the *ability to ensure fitness for purpose for all aspects of the problem including production, operation, maintenance and disposal*. While this statement echoes our thinking, it is included in the design section and effectively refers to multi-objective design (Design for X). It can therefore be argued that UK-SPEC does not fully recognize that all engineers need to know how to implement their designs in the form of physical products or systems. In addition, UK-SPEC makes no specific mention of the operational phase of the product or system lifecycle, apart from the need to promote sustainable development.

From the foregoing discussion, it is apparent that the CDIO Syllabus has many advantages over UK-SPEC, the most obvious of which are the following:

- Although UK-SPEC contains more learning outcomes than the ABET criteria, it still lacks the fine detail of the Syllabus.
- UK-SPEC is not self-contained, *i.e.*, it delegates responsibility for important personal and interpersonal skills to the Qualifications and Curriculum Authority, which provides guidance on a limited set of skills that are not specifically those required by engineering graduates.
- There is limited coverage of professional skills in UK-SPEC compared with the CDIO Syllabus. Although formal training in employment, which used to be widespread in the United Kingdom, is now rare, employers expect graduates to have at least some of the professional skills needed to move directly into positions of responsibility.
- UK-SPEC does not reflect the need for competence in all aspects of the product or system lifecycle. The implementation and operational phases, in particular, are not addressed by appropriate learning outcomes.

It may be argued that UK-SPEC is less helpful than it could be because it limits itself to listing a series of learning outcomes. Although topic-based, the Syllabus is supported by a rationale, a set of standards and a process for developing program-specific learning outcomes. The comprehensive nature of the Syllabus also means that it will invariably cover any learning outcomes required for accreditation purposes. However, when combined with its other elements, a CDIO approach achieves a more fundamental objective, which is to indicate clearly how an engineering program can be improved, and not simply accredited.

– P. Armstrong, Queen's University Belfast

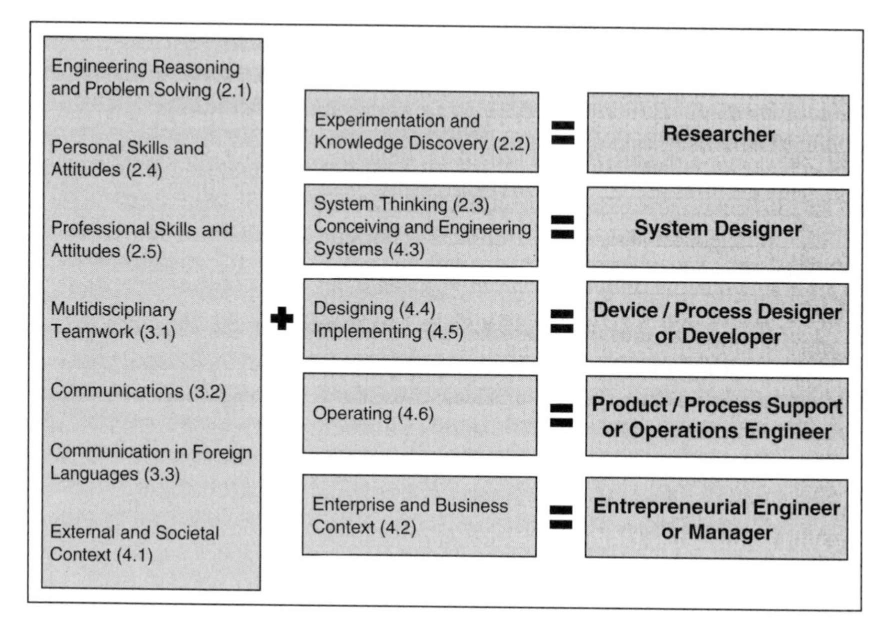

FIGURE 3.6. **PROFESSIONAL ENGINEERING CAREER TRACKS IMPLICITLY IDENTIFIED IN THE CDIO SYLLABUS**

As an independent check on the comprehensiveness of the CDIO Syllabus, it was compared with generic skills needed by engineers in five different career tracks. The generic skills applicable to all tracks include: 2.1 *Engineering Reasoning and Problem Solving*, 2.4 *Personal Skills and Attitudes*, 2.5 *Professional Skills and Attitudes*, 3.1 *Multidisciplinary Teamwork*, 3.2 *Communications*, 3.3 *Communications in a Foreign Language*, and 4.1 *External and Societal Context*. There are at least five professional tracks that engineers can and do follow, according to their individual talents and interests. The tracks, and sections of the Syllabus that support them, are shown in Figure 3.6.

Of course, no graduating engineer will be expert in all of these potential tracks, and in fact, may not be expert in any. However, the paradigm of modern engineering practice is that an individual's role will change and evolve. The graduating engineer must be able to interact in an informed way with individuals in each of these tracks, and must be educated as a generalist, prepared to follow a career that leads to any one or any combination of these tracks.

Contemporary themes in engineering – innovation and sustainability

The CDIO Syllabus was written with the intent of creating a long-lived, stable enumeration of the knowledge and skills required of an engineer. At the same time, contemporary themes emerge in importance for both engineering and engineering education. Currently, the concepts of *innovation* and *sustainability,* or *sustainable development,* fall in this category. Such themes do not appear explicitly in the higher-level organization of the Syllabus, but we believe that the constituent knowledge and skills that support these themes are, in fact, present.

An excellent guide for the teaching of engineering for sustainable development was recently produced by the Royal Academy of Engineering in the United Kingdom [10]. It cites the often-quoted definition of sustainability from the 1987 Report of the United Nations World Commission on Environment and Development (The Bruntland Report): "Humanity has the ability to make development sustainable—to ensure that it meets the needs of the present without compromising the ability of future generations to meet their own needs." The Syllabus contains one explicit reference to sustainability, in Section 4.4.6 *Multiobjective Design*, where design for environmental sustainability is among the objects for the design of products, processes, or systems. The Royal Academy report goes on to describe sustainability as the constructive intersection of technocentric, sociocentric, and ecocentric concerns. A scan of the Syllabus for these three categories of concerns reveals four third-level (X.X.X) topics that discuss issues of appropriate development and use of technology, five third-level topics that address various distinct environmental issues, and nine third-level topics that address distinct aspects of an engineer's responsibility to consider society, societal issues in design, and society's regulation of engineering. The Royal Academy report defines twelve guiding principles of engineering for sustainable development. Table 3.4 compares these twelve principles with the closest corresponding topics of the

TABLE 3.4. PRINCIPLES OF SUSTAINABILITY COMPARED WITH THE CDIO SYLLABUS

	Sustainability Principle	The CDIO Syllabus	
1	Look beyond your own locality and the immediate future	4.1.1	Rules and Responsibilities of Engineers
		4.1.2	The Impact of Engineering in Society
		4.1.6	Developing a Global Perspective
2	Innovate and be creative	2.4.3	Creative Thinking
3	Seek a balanced solution	2.3.4	Trade-offs, Judgment and Balance in Resolution
4	Seek engagement from all stakeholders	4.3.1	Setting System Goals and Requirements
5	Make sure you know the needs and wants		
6	Plan and manage effectively	4.3.4	Development Project Management
7	Give sustainability the benefit of any doubt	4.4.6	Multi-objective Design
8	If polluters must pollute, then they must pay as well		
9	Adopt a holistic 'cradle-to-grave' approach	2.3.1	Thinking Holistically
10	Do things right, having decided on the right thing to do	2.5.1	Professional Ethics, Integrity, Responsibility, and Accountability
11	Beware cost cutting that masquerades as value engineering	2.1.5	Solution and Recommendation
		2.4.4	Critical Thinking
12	Practice what you preach	2.5.1	Professional Ethics, Integrity, Responsibility, and Accountability

Syllabus. In almost all cases, the Syllabus contains the skills or knowledge associated with the principle or a slight generalization of it.

In its recent Innovation Survey, the Confederation of British Industry (CBI) defines innovation broadly as "the successful exploitation of new ideas" [11]. From this and similar definitions, the Cambridge-MIT Institute has defined the set of knowledge, skills and attitudes, which lay the foundation for innovation [12]. In brief, these include a deep conceptual understanding of fundamentals, the skills to exploit ideas, and a sense of self-empowerment from learning. Said another way, these are the knowledge to innovate, the skills to innovate, and a positive attitude towards taking the risks necessary for innovation. When comparing this model with the CDIO Syllabus, we find the knowledge of a technical discipline listed in Section 1, reinforced by the goal of developing a deeper working knowledge of the technical fundamentals. The skills required to exploit ideas include understanding the needs of the customer (CDIO Syllabus 4.3.1), application of appropriate technology (4.3.2), communications (3.2 and 3.3), and teamwork (3.1). The character traits that underlie an inclination to innovate include a willingness to take risk (2.4.1), perseverance (2.4.2) and creative thinking (2.4.3). In addition a successful innovator understands the enterprise (4.2.2), working within established organizations (4.2.3), and entrepreneurial concerns (4.2.4).

These two examples illustrate that the underlying knowledge, skills, and attitudes of education in the contemporary themes of sustainability and innovation are present in the CDIO Syllabus, even though the headings are not.

Generalizing the CDIO Syllabus

We have created the CDIO Syllabus so that, in principle, it is applicable to any field of engineering. As mentioned above, we have chosen words, such as implement, which are recognizable to all engineers, although when customizing the Syllabus, a civil engineer might substitute constructing, while a software engineer might prefer coding and testing. The intention was to create terms at the second, or X.X, level that are completely generic to all forms of engineering, and to try to do so as much as possible at the third, or X.X.X, level. At the lowest level, some unavoidable references to engineering sectors that produce discrete objects, such as cars, aircraft, and electronic devices, may have occurred. However, one could readily modify these lowest-level topics and outcomes with descriptors that are more appropriate to chemical, biological, software, or other forms of engineering.

At the highest level, the Syllabus could be abstracted to a statement of educational goals for virtually any university education. Section 1 would generalize from *Technical Knowledge and Reasoning* to Disciplinary Knowledge and Reasoning, a small step. Sections 2 and 3, *Personal and Professional Skills and Attitudes* and *Interpersonal Skills*, would remain largely unchanged. The sole exception is that the relative organization of Section 2 might change to emphasize other modes of thought. Section 4, *Conceiving, Designing, Implementing and Operating Systems in an Enterprise and Societal Context*, is the most

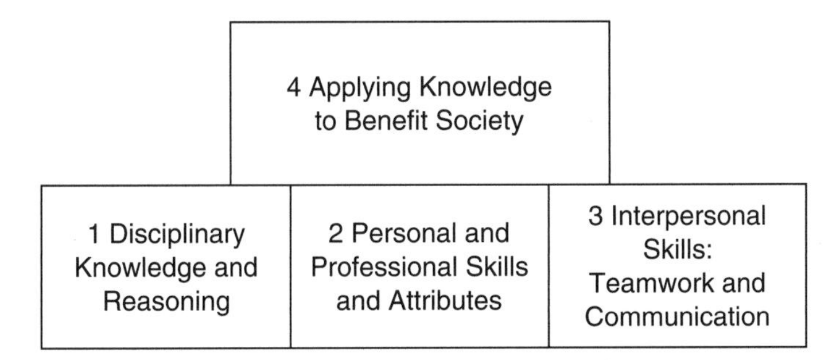

FIGURE 3.7. **HIGH-LEVEL ORGANIZATION OF THE GENERALIZED SYLLABUS**

challenging to generalize. If one abstracts this topic to "how engineers add value to society," then the generalization of Section 4 would be Applying Knowledge to Benefit Society. For example, in law education, this section would include the professional skills of law practice, as opposed to knowledge of the law, which would be in Section 1. Likewise in medical education, Section 4 would include the professional practice of medicine, as opposed to the knowledge of medicine, which would be in Section 1. Figure 3.7 shows the high-level organization of the generalized Syllabus, in parallel with Figure 3.2, which is the high-level organization specialized for engineering.

LEARNING OUTCOMES AND STUDENT PROFICIENCY LEVELS

The CDIO Syllabus is a detailed list of knowledge and skills in which a graduating engineer should have developed some level of proficiency. It comprehensively addresses the first part of the central question before us in this chapter:

> *What is the full set of knowledge, skills, and attitudes that engineering students should possess as they leave the university, and at what level of proficiency?*

However, Standard 2, given above, calls for more than a mere listing of topics. It requires that a program set *specific, detailed learning outcomes for personal and interpersonal skills; and product, process, and system building skills consistent with program goals and validated by program stakeholders*. In order to translate this list of knowledge, skills, and attitudes into learning outcomes, detailed levels of expected proficiency need to be established for all of the topics in the Syllabus.

The recommended process for establishing proficiency levels and learning outcomes is as follows:

- Review the generic CDIO Syllabus, as shown in Table 3.1, and make modifications or additions to customize it for a specific course of study within the technical and national context of the program.

- Identify the important stakeholders of the program—both internal and external to the university. It is highly desirable to capture the opinions of representative stakeholders of the educational program, and to encourage consensus of both individual viewpoints and collective wisdom. This is the intent of the phrase *validated by program stakeholders* in Standard 2.
- Determine a means of engaging stakeholders and summarizing their opinions. The approach most used to date in our programs is a survey that collects data from stakeholders.
- Faculty discussions help to interpret the results of stakeholder input, and these discussions lead to consensus on expected levels of proficiency.
- These expected levels of proficiency can then be translated into more formally stated learning outcomes that are the basis for instructional design and student learning assessment.

The result of this process is the answer to the second part of the question, "*. . . at what level of proficiency?*"

Two examples of this process for determining proficiency levels for the CDIO Syllabus are described below. In the first example, the Syllabus, as first developed, was used as the basis of the survey of program stakeholders. This general example is based on the experiences of programs in the United States and Sweden [13]-[14], and focuses on Sections 2, 3, and 4 in the Syllabus. In the second example, program faculty decided to augment their survey with topics included in Section 1, *Technical Knowledge and Reasoning,* in order to develop a more complete view of stakeholder opinions on learning outcomes [15].

Learning outcome studies by the four founding universities

Early in The CDIO Initiative, Chalmers University of Technology (Chalmers), Linköping University (LiU), the Royal Institute of Technology (KTH), and the Massachusetts Institute of Technology (MIT) conducted parallel studies to set the expected levels of proficiency in the topics of the CDIO Syllabus with their respective program stakeholders [14]. To allow more direct comparison of the results, the four universities agreed to use the generic form of the Syllabus without alteration (although the Swedish universities translated the document into Swedish).

Engineering education has many stakeholder communities who might be included in survey and consensus processes. These groups include faculty who are internal and external to the target program, alumni of various ages, industry representatives, and peers at other universities. Advisory boards, administrators, and faculty in other departments can also be included. Depending on local culture, engineering students can be surveyed as well.

In the studies conducted by the four founding universities, we included the same six groups of stakeholders:

- university faculty
- mid- to upper-level industry leaders

- alumni about five years past graduation
- alumni about 15 years past graduation
- first-year students
- final-year students

We chose alumni groups who were young enough to recall their education in some detail, yet old enough to be able to reflect on it with meaningful perspective. We selected the two groups to determine whether the opinions of alumni changed over time. We included first-year and final-year students to gauge changes in their expectations with experience and to correlate their opinions with those of the other groups.

Survey process for determining expected proficiency levels

There are several alternatives for gathering stakeholder input, including interviews and focus groups, but the most common one used in our programs has been a survey. In the example of the four founding programs, the survey process used a questionnaire to collect data from stakeholders about the expected proficiency levels of CDIO Syllabus knowledge and skills. The survey instrument asked questions in such a way that information was collected for each item in the Syllabus at the second, or X.X, level of detail, and at the third, or X.X.X, level of detail. (See Table 3.1 for the Syllabus at the second and third levels of detail.) Both quantitative and qualitative responses were solicited. Respondents were given a set of definitions to ensure reasonable consistency of interpretation and increase the reliability of the responses. Respondents were also given the complete CDIO Syllabus (See Appendix A) and background reading on the program. A representative sample of 20 to 30 respondents usually captured all of the important trends in stakeholder opinions.

For each second-level (X.X-level) Syllabus topic, respondents were asked to indicate an expected proficiency level using a 5-point scale. Table 3.5 shows the rating scale. Scale points designate absolute level of competence expected in the activities or experiences of engineers. They are not relative measures of skills compared with other graduating engineers. For example, *5 To be able to lead or innovate in* requires a level of proficiency attained by experts in a particular discipline or area. In addition, respondents were encouraged to include brief statements elaborating on their ratings.

TABLE 3.5. **EXPECTED LEVEL OF PROFICIENCY IN CDIO SYLLABUS KNOWLEDGE AND SKILLS**

1	To have experienced or been exposed to
2	To be able to participate in and contribute to
3	To be able to understand and explain
4	To be skilled in the practice or implementation of
5	To be able to lead or innovate in

TABLE 3.6. SAMPLE RESPONSE TO EXPECTED PROFICIENCY QUESTIONNAIRE

2.1	Engineering and Problem Solving	4
2.1.1	Problem Identification and Framing	+
2.1.2	Modeling	–
2.1.3	Estimation and Qualitative Analysis	
2.1.4	Analysis With Uncertainty	–
2.1.5	Closing the Problem	+

For each Syllabus topic at the second (X.X) level, respondents were asked to choose one or two topics at the third (X.X.X) level of detail in which students should develop relatively higher proficiency (+), and one or two topics in which they could develop relatively lower proficiency (–). Respondents were encouraged, though not required, to balance the plusses and minuses within any X.X group. Table 3.6 illustrates a sample response to one Syllabus topic.

Qualitative and quantitative data were gathered on the 14 second-level and the 67 third-level Syllabus topics from respondents in each of the stakeholder groups. For each of the four programs, mean responses for each of the stakeholder groups were calculated. Statistical tests, for example, Student's t-test, determined whether differences in the means were meaningful. The qualitative comments of the respondents were examined to determine if they led to any generalizations of the understanding of the trends and differences among different stakeholder groups.

Survey results at MIT

Figure 3.8 shows the results of the proficiency-level surveys at MIT. Four stakeholder groups are included: faculty, industry leaders, and two groups of alumni. These groups are referred to as *MIT Professionals* in the survey. An examination of Figure 3.8 reveals that in the comparisons of expected proficiency, *Engineering Reasoning and Problem Solving* (2.1), *Communications* (3.2), *Designing* (4.4), and *Personal Skills and Attitudes* (2.4), with proficiency levels between 3.4 and 4, are the most highly ranked topics. Experts consistently cite these four topics as among the most important skills of engineering, and their high ranking is not a surprise. These means correspond to a scale rating of *4 To be skilled in the practice or implementation of* these topics.

External and Societal Context (4.1), *Enterprise and Business Context* (4.2), *Implementing* (4.5) and *Operating* (4.6) are rated low, with expected proficiencies near a rating of *2 To be able to participate in and contribute to*. The lower ratings in the first two topics could not be explained through respondents' comments. Respondents, however, specifically noted that the lower ratings on *Implementing* (4.5) and *Operating* (4.6) relate to their suggestions that these topics may be better learned on the job or may be too domain specific to teach at a university. One of the Syllabus topics, *Communications in Foreign Languages* (3.3), was not included in the surveys conducted by MIT.

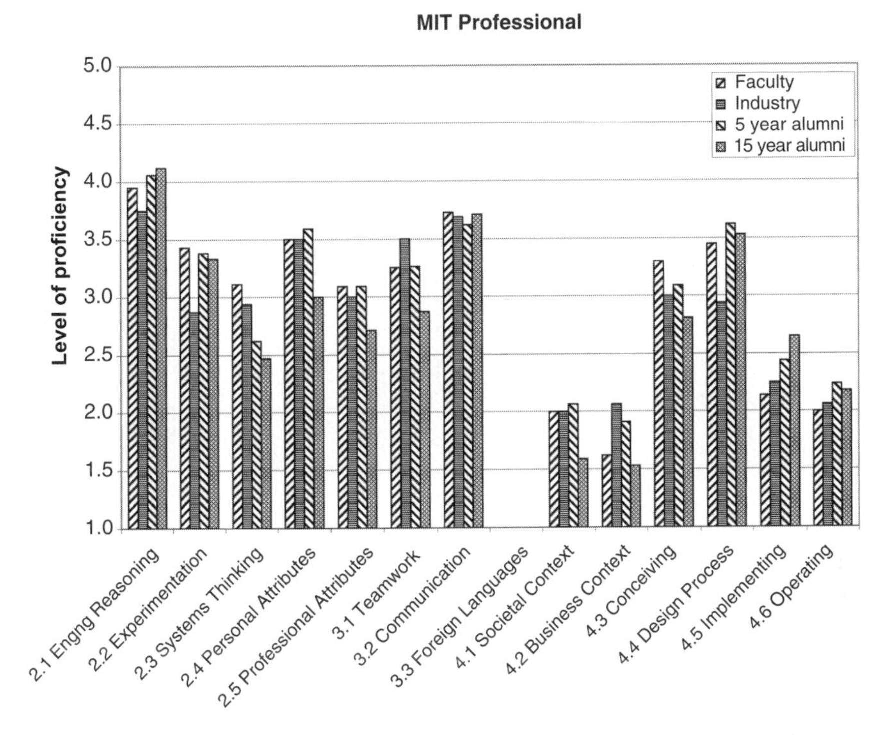

FIGURE 3.8. **EXPECTED PROFICIENCY LEVELS REPORTED BY STAKEHOLDER GROUPS AT MIT**

The most striking outcome of the MIT survey is the statistical agreement on all but one topic among the four professional stakeholder groups. This result was totally unexpected. It indicates that when asked about a specific program, with a well-defined enumeration of skills and a clear set of rubrics, different stakeholders will have very similar views with regard to expected levels of proficiency. This convergence is a powerful starting point for curriculum design, and an important benchmark for student learning assessment.

Survey results at three Swedish Universities

Figure 3.9 shows the results of the stakeholder surveys for the three Swedish universities: Chalmers, KTH, and LiU. For each Syllabus topic, the mean rating of professional groups at the three universities is shown. The topics most highly rated in the Swedish surveys of expected proficiencies are *Engineering Reasoning and Problem Solving* (2.1), *System Thinking* (2.3), *Personal Skills and Attitudes* (2.4), *Teamwork* (3.1), *Communications* (3.2), and *Communications in Foreign Languages* (3.3), all with mean proficiency ratings near the scale rating

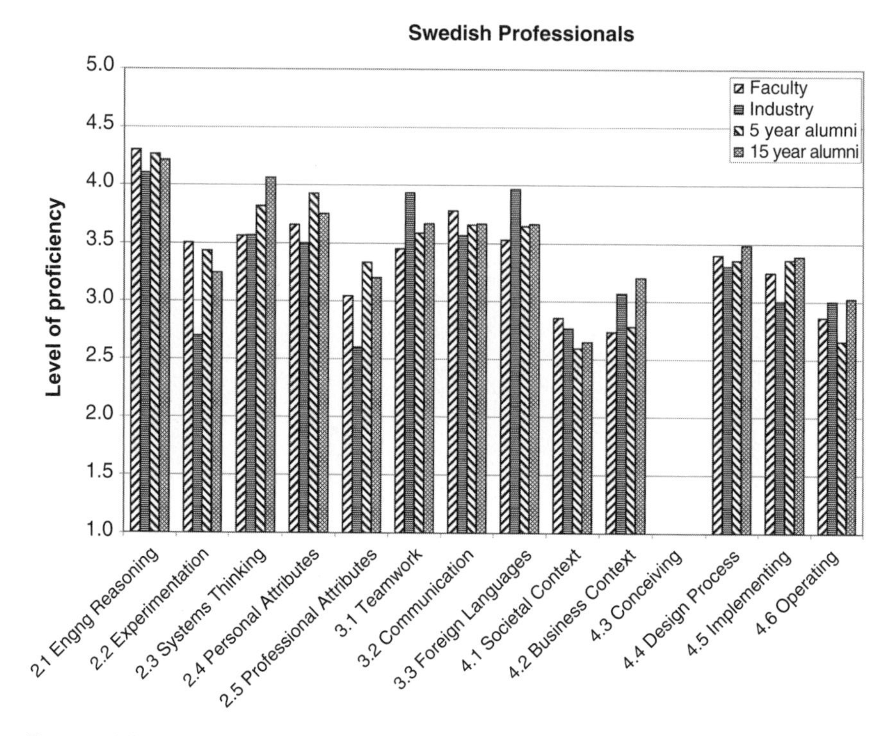

FIGURE 3.9. **EXPECTED PROFICIENCY LEVELS REPORTED BY STAKEHOLDER GROUPS AT THREE SWEDISH UNIVERSITIES**

of *4 To be skilled in the practice or implementation of*. Comparisons between the MIT results and those of the Swedish universities are influenced by differences in English- and Swedish-language versions of the CDIO Syllabus. *External and Societal Context* (4.1), *Enterprise and Business Context* (4.2), and *Operating* (4.6) were rated relatively low in the Swedish surveys, with mean proficiency ratings near the scale rating of *3 To be able to understand and explain*. The ratings are not as low as those in the MIT survey. Results of *4.3 Conceiving* are not included because of a software error in the data analysis phase.

Comparisons across all four Universities

Ratings from the professional stakeholder groups (faculty, industry leaders, and two groups of alumni) were compared across MIT, Chalmers, KTH, and LiU. Figure 3.10 shows that the agreement among professional engineers in industry across the universities is, in general, very good for all topics except for *Systems Thinking* (2.3) where MIT respondents rated the expected proficiency level between scale ratings of 2 and 3, while LiU respondents rated the expected proficiency level above 4.

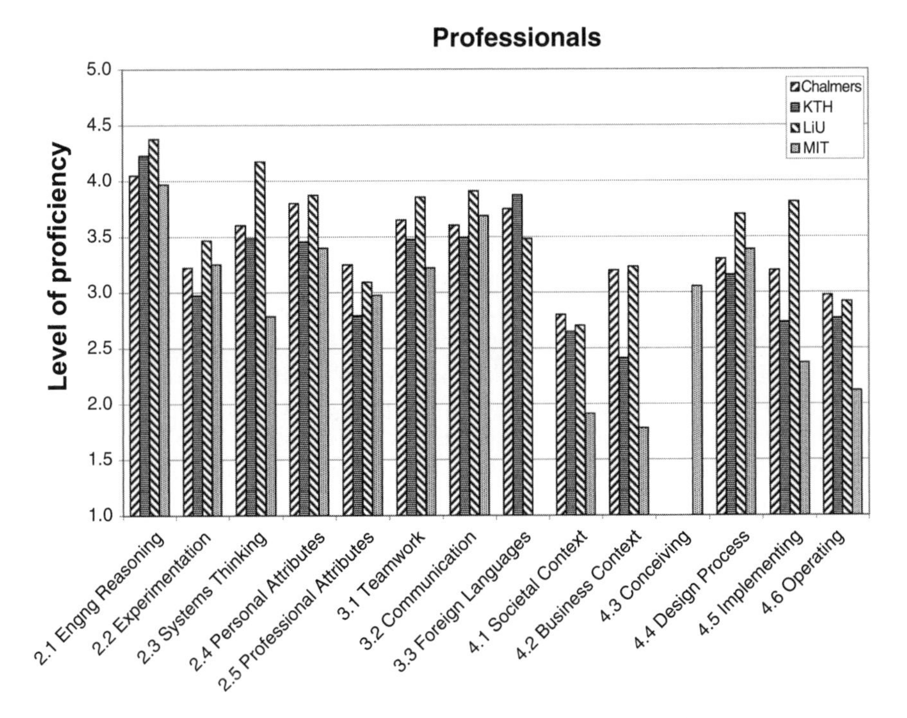

FIGURE 3.10. AVERAGE OF EXPECTED PROFICIENCY LEVELS REPORTED BY THE PROFESSIONAL STAKEHOLDER GROUPS AT FOUR UNIVERSITIES

Major disagreements also occur for *External and Societal Context* (4.1), *Enterprise and Business Context* (4.2), *Implementing* (4.5) and *Operating* (4.6). Statistical analyses reveal some differences in the ratings among the Swedish stakeholder groups, particularly for *Enterprise and Business Context* (4.2) and *Implementing* (4.5), but the most significant difference is the considerably lower level of proficiency expected for topics 4.1, 4.2, 4.5 and 4.6 by MIT respondents compared with the Swedish respondent groups. One possible explanation is that programs at the Swedish universities are 4.5 years, while the MIT program is four years. Another explanation is that these skills are emphasized in the required diploma thesis project that most Swedish engineering students complete in industrial settings under faculty supervision. It should be noted that Swedish stakeholder groups represented a wider range of engineering disciplines, for example, vehicle engineering, mechanical engineering, and applied physics and electrical engineering, while MIT respondents were primarily aerospace professionals.

The most significant result of the surveys is the similarity of opinion among each university's respective stakeholder groups. This degree of consensus was unexpected, and helps to validate expected levels of proficiency in knowledge and skills for students graduating from CDIO programs.

Learning outcome studies at Queen's University Belfast

The School of Mechanical and Manufacturing Engineering at Queen's University Belfast (QUB) conducted a similar survey of expected proficiency levels. As with the surveys at MIT and the three Swedish universities, the focus was on Sections 2, 3 and 4 of the Syllabus. The most important category of stakeholders for the QUB survey was QUB alumni. As graduates, they are familiar with QUB's programs, and, as experienced professionals, they know how well the knowledge and skills acquired during their studies prepared them for their careers. About 800 alumni, who had graduated between 5 and 30 years ago, were invited to participate in the survey.

In addition to questions about the CDIO Syllabus, the survey asked alumni about other curricular issues, for example, appropriate topics in mathematics, depth of learning in the engineering sciences, and additional subjects included in the program. As with most universities in the United Kingdom, Queen's University supplements its mechanical engineering science and mathematics courses with courses in design, management, economics, law, electronics, and computer programming. It is generally difficult to gauge which subjects to include and how much curriculum time to devote to these additional subjects. As a further objective of the survey, Queen's University wanted confirmation that alumni supported the curriculum changes envisaged in the transition to the CDIO approach. Finally, it was felt that basic information on the careers pursued by graduates would be useful reference material for future debate. In the end, asking for all this information required a fairly lengthy questionnaire.

Unlike the previous surveys at MIT, Chalmers, KTH, and LiU, Queen's University did not use the 5-point scale, found in Table 3.4, to ask about expected levels of proficiency in the knowledge and skills of the Syllabus. Instead, a 5-point level of importance scale was used. The scale ranged from *1 Of no importance* to *5 Essential*. The scale was changed mainly because the importance scale was more suitable to a greater number of items in their questionnaire. The correspondence between the two scales is a matter for debate, but if a particular skill is more important than another, it can probably be assumed that a higher level of proficiency is required.

About 200 hundred alumni responded—a strong response rate for a postal survey. Means were calculated for the items related to Sections 2, 3, and 4 of the Syllabus, for comparison with results from alumni at MIT. Figure 3.11 shows the comparison of mean responses from QUB alumni with MIT alumni (from the study described in this chapter). The agreement found between the MIT and Swedish results is repeated in the comparison of QUB and MIT. (This assumes correspondence between the proficiency and importance scales. The similarity of the results suggests that there is acceptable correspondence.) A close examination does, however, reveal some interesting differences. *Enterprise and Business Context* (4.2) is rated significantly higher by the QUB alumni, compared with their MIT counterparts. This is undoubtedly a reflection of the fact that a relatively high percentage of QUB

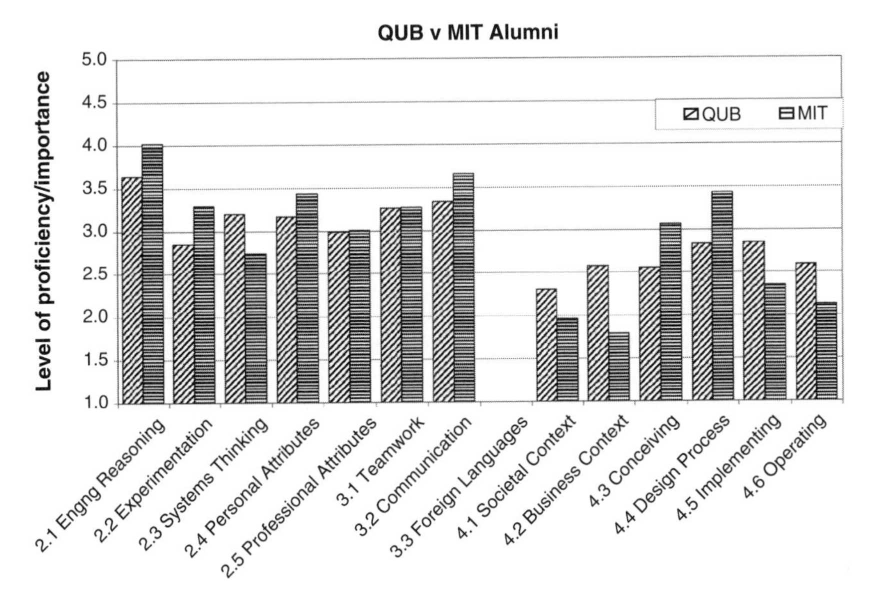

FIGURE 3.11. COMPARISON OF RESULTS FOR QUB AND MIT ALUMNI (3.3 NOT INCLUDED IN THE SURVEY)

graduates are employed in small companies that dominate the local economy. In such companies, professional engineers tend to be involved in the general running of the company, and have to take on management and financial responsibilities. Hence, grounding in management and business skills at university is more important than it would be for graduates employed by larger companies. A further difference between the MIT and QUB alumni is the relative importance placed on design and implementation. As Figure 3.11 indicates, MIT alumni felt that the proficiency level in topics related to implementation should be much lower than the proficiency level in design. In contrast, QUB alumni thought that implementation was just as important as design. Respondents of the Swedish surveys also rated implementation more highly than did the MIT respondents, but this may simply reflect differences among the engineering disciplines involved in the MIT, QUB, and Swedish surveys. Aerospace graduates from MIT are probably less likely than the others to work in areas related to manufacturing and tend to regard implementation as less important than design.

The differences in the details of the survey results provide a justification for each university to conduct its own survey. Each university then can adjust the emphasis placed on each section of the Syllabus to match the specific needs of its graduates. It is also worth noting that the views of Queen's University students and faculty on Sections 2, 3, and 4 of the Syllabus differed from those of the alumni, and while similar, students and faculty results showed interesting differences from each other. Exposing misconceptions among students and faculty is a further benefit of each university conducting its own surveys.

In the section of the alumni questionnaire on additional subjects, twenty subjects were listed. Respondents were asked to rate each one using the five-point scale of importance. The results indicated that alumni regarded manufacturing, management, and business courses as particularly important. Control received a lower rating, which may be a reflection on the theoretical nature of control courses in universities. However, the lowest rating was given to computer programming, a result that was of particular interest to the School since some faculty questioned the need to teach computer programming. Their argument was that in view of the range of software applications now available, mechanical and manufacturing engineers are unlikely to need programming skills. The views of the alumni were taken into account and after further debate, computer programming was removed from the curriculum.

Part of the alumni questionnaire dealt with the teaching of engineering science subjects. For each of the main engineering science subjects taught, respondents were asked to rate the relative importance of understanding basic principles; being able to define specific relationships between variables; and being able to express these relationships mathematically. The results indicated that a high percentage (more than 80%) felt that familiarity with basic principles is very important or essential, while relatively few (about 30%) felt the same way about the formation and application of mathematical relationships.

Another section of the questionnaire dealt with the balance required between different areas of the curriculum. Alumni were asked how much time *should* be devoted to each area, compared with their experience as QUB students. The rating scale adopted in this case ranged from *1 Considerably less time* to *5 Considerably more time*. Table 3.7 shows the ranking of each area on the basis of the average rating recorded. In addition, the percentage of respondents requesting more or considerably more time is shown separately for younger alumni (fewer than 10 years past graduation), and older alumni (more than 20 years past graduation). Despite the difference between the two groups in terms of age and experience, there is strong agreement that

TABLE 3.7. **TIME TO BE ALLOCATED TO EACH AREA OF THE CURRICULUM**

			% More or Considerably More Time	
Rank	Area of Curriculum	Ave. Rating	Younger Alumni	Older Alumni
5	Mechanical Engineering Science	3.06	18	9
6	Mathematics	2.35	5	0
3	Additional Subjects	2.44	49	48
4	Practical Work: Laboratory Classes	3.13	26	21
1	Practical Work: Design & Build	3.84	64	71
2	Developing Skills & Attributes	3.81	67	59

more time should be devoted to three areas: practical work in the form of design-and-build exercises: the development of skills and attributes: and additional subjects beyond engineering science and mathematics.

Overall, the QUB surveys were highly valuable in the course of curriculum design. The results obtained for Sections 2, 3, and 4 of the Syllabus provided useful input on the emphasis to be placed on each area of knowledge, skills, and attitudes. In addition, results highlighted areas where faculty and students have opinions that do not correspond to alumni's more authoritative views. The extended questionnaire also assisted decision making on topics to be covered in mathematics and the additional subjects to be included in the curriculum. However, a striking result of the survey is the support it provides for a CDIO approach to engineering education. QUB graduates regard an understanding of fundamental principles as very important, and fully support the allocation of additional time to design-build experiences and the development of skills and attitudes. This result is a timely endorsement of the university's participation in the CDIO Initiative.

Interpreting expected levels of proficiency as learning outcomes

Having determined the expected level of proficiency for each CDIO Syllabus topic at the second and third levels of detail, the remaining task is to formulate corresponding learning outcomes. This formulation requires three steps:

- Choosing a taxonomy of learning outcomes.
- Developing a correspondence between the taxonomy and the rating scale used to determine expected levels of proficiency.
- Specifying a learning outcome for each of the most detailed topics in the Syllabus, corresponding to the taxonomy and the appropriate proficiency rating.

The first step in formulating learning outcomes is to choose an appropriate taxonomy. From among several possibilities, the one most widely used by our programs is that of Bloom and his colleagues [15]-[17]. Briefly, Bloom's taxonomy divides learning into three potentially overlapping domains. The cognitive domain addresses knowledge and reasoning; the affective domain includes attitudes and values; the psychomotor domain describes skills requiring mobility and manipulation. Each of the three domains is classified into five or six hierarchical levels.

In order to specify learning outcomes derived from the Syllabus, a correspondence must be developed between Bloom's taxonomy and the rating scale used to determine expected levels of proficiency. Table 3.8 illustrates such a correspondence. For example, in Bloom's cognitive domain, there is no skill that corresponds with a rating of *1 To have experienced or been exposed to*. Looking further, however, a rating of *2 To be able to participate in and contribute to corresponds* to Knowledge; a rating of *3 To be able to understand*

TABLE 3.8. CORRESPONDENCE OF PROFICIENCY RATING SCALE AND BLOOM'S TAXONOMY

Proficiency rating scale – stakeholder surveys	Bloom's taxonomy - cognitive domain	Examples of learning outcomes based on the CDIO syllabus
1 To have experience or been exposed to		
2 To be able to participate n and contribute to	Knowledge	List assumptions and sources of bias
3 To be able to understand and explain	Comprehension	Explain discrepancies in results
4 To be skilled in the practice or implementation of	Application	Practice engineering cost-benefit and risk analysis
	Analysis	Discriminate hypotheses to be tested
5 To be able to lead or innovate	Synthesis	Construct the abstractions necessary to model the system
	Evaluation	Make reasonable judgments about supporting evidence

and explain is associated with Comprehension; a rating of *4 To be skilled in the practice or implementation of* maps to two levels, Application and Analysis; and finally, a rating of *5 To be able to lead or innovate* corresponds to Bloom's highest cognitive levels of Synthesis and Evaluation. Similar approximate correspondences can be drawn to the affective and psychomotor domains.

The final step in converting Syllabus topics into specific learning outcomes is to associate each topic phrase with a verb that best describes the level of proficiency determined by program stakeholders. Each level of Bloom's taxonomy can be expressed with specific verbs. For example, Synthesis in the cognitive domain includes such skills as *formulate, create, construct,* and *reorganize*. Table 3.8 gives examples of specific learning outcomes derived from the Syllabus presented at the appropriate levels of expected proficiency. While it may have been possible to specify program learning outcomes without having gotten stakeholder input, the rigorous survey process used to set expected levels of proficiency enabled us to set more realistic learning outcomes for engineering students.

SUMMARY

This chapter focused on defining the CDIO Syllabus, describing its structure and development, and showing how the Syllabus can be used as the basis for determining stakeholder consensus on expected learning outcomes. It is a generalized statement of goals for engineering education that flows directly from the actual roles of engineers. It is comprehensive in that it includes all of the knowledge, skills, and attitudes expected of a graduating engineer.

Although the Syllabus is a generalized statement, it can be customized to meet local program needs. The process of customization includes the definition of disciplinary content for Section *1 Technical Reasoning and Problem Solving* and adjustments to the rest of the Syllabus, particularly at the third- and fourth-levels of detail. Surveys gather input from program stakeholders about the expected levels of proficiency, or importance, of each Syllabus topic. Results provide guidance for curriculum design and learning assessment.

Surveys conducted by representative programs yielded some interesting results. The agreement among the faculty, industry leaders, and alumni on the expected levels of proficiency of graduating engineers was significant and unexpected. Surveys indicated that the skills for which the proficiency expectations are the highest include engineering reasoning, personal attributes, communications, and design. These four skills are consistently among those cited as most important in a graduating engineer. In programs at universities in Sweden, the expected proficiency in foreign language communication is also high.

The CDIO Syllabus, customized with results of stakeholder surveys, lays the foundation for specific learning outcomes, curriculum planning and integration, teaching and learning practice, and outcomes-based assessment. The process of integrating the Syllabus into a program's curriculum is the subject of Chapter Four. Approaches to teaching and learning the content of the Syllabus are described in Chapter Six. Student assessment of these learning outcomes is the focus of Chapter Seven.

DISCUSSION QUESTIONS

1. How would you rate your own proficiency in the knowledge and skills of the CDIO Syllabus?
2. How would you rate the proficiency levels expected of graduates of your program?
3. How do your ratings compare with those of the programs described in this chapter?
4. In what ways can you implement the suggested processes to define your program learning outcomes and validate them with your program stakeholders?

References

[1] King, W. J., "The Unwritten Laws of Engineering", *Mechanical Engineering*, May 1944, pp. 323-326; June 1944, pp. 398-402; July 1944, pp. 459-462.
[2] Gordon, B. M., "What is an Engineer?", Invited Keynote Presentation, Annual Conference of the European Society for Engineering, University of Erlangen-Nürnberg, 1984.

[3] The Boeing Company, "Desired Attributes of an Engineer: Participation with Universities", 1996. Available at http://www.boeing.com/companyoffices/pwu/attributes/attributes.html
[4] Augustine, N. R., "Socioengineering (and Augustine's Second Law Thereof)", *The Bridge*, Fall 1994, pp. 3-14.
[5] World Chemical Engineering Council, "How Does Chemical Engineering Education Meet the Requirements of Employment?" 2004.
[6] Accreditation Board of Engineering and Technology, "Criteria for Accrediting Engineering Programs: Effective for Evaluations During the 2000-2001 Accreditation Cycle", 2000. Available at http://www.abet.org
[7] Massachusetts Institute of Technology, School of Engineering Committee on Engineering Undergraduate Education, "Eight Goals of an Undergraduate Education", Cambridge, MA, 1988. Unpublished internal document.
[8] Massachusetts Institute of Technology, Task Force on Student Life and Learning, *Task Force Report, 22 April 1998*, Cambridge, MA, 1988. Unpublished internal document.
[9] Engineering Council, "UK Standards for Professional Engineering Competence: The Accreditation of Higher Education Programs", 2004. Available at http://www.iee.org/professionalregistration/ukspec.cfm
[10] Dodds, R., and Venables, R., *Engineering for Sustainable Development: Guiding Principles*, The Royal Academy of Engineering, London, 2005.
[11] The Confederation of British Industry (CBI), *Innovation Survey 2005*, London, Author, 2005.
[12] Cambridge-Massachusetts Institute of Technology Institute (CMI).
[13] Crawley, E. F., *The CDIO Syllabus – A Statement of Goals for Undergraduate Engineering Education*, Massachusetts Institute of Technology, Department of Aeronautics and Astronautics, Cambridge, Massachusetts, 2001.
[14] Bankel, J., Berggren, K.-F., Blom, K., Crawley, E. F., Wiklund, I., and Östlund, S., "The CDIO Syllabus – A Comparative Study of Expected Student Proficiency", *European Journal of Engineering Education*, Vol. 28, No. 3, 2003, pp. 297-315.
[15] Bloom, B. S., Englehatt, M. D., Furst, E. J., Hill, W. H., and Krathwohl, D. R., *Taxonomy of Educational Objectives: Handbook I – Cognitive Domain*, McKay, New York, 1956.
[16] Krathwohl, D. R., Bloom, B. S., and Masia, B. B., *Taxonomy of Educational Objectives: Handbook II – Affective Domain,* McKay, New York, 1964.
[17] Simpson, E. J., *The Classification of Educational Objectives in the Psychomotor Domain,* Gryphon House, Washington, DC, 1972.

CHAPTER FOUR INTEGRATED CURRICULUM DESIGN

WITH K. EDSTRÖM, S. GUNNARSSON, AND G. GUSTAFSSON

INTRODUCTION

We have now reached a transition point in our discussion. In Chapter Two, we posed the two central questions that any approach to improving engineering education must address:

- *What is the full set of knowledge, skills, and attitudes that engineering students should possess as they leave the university, and at what level of proficiency?*
- *How can we do better at ensuring that students learn these skills?*

As discussed in the previous chapters, there are compelling reasons for university engineering programs to educate students in a broad set of personal and interpersonal skills, and product, process, and system building skills, as well as to instruct them in the technical disciplines. We argued that the best way to accomplish this is to stress the fundamentals, and to set the education in the context of conceiving-designing-implementing-operating products, processes, and systems (the essence of CDIO Standard 1); that students are expected to achieve a comprehensive set of learning outcomes, as defined by the CDIO Syllabus; and that learning outcomes should be comprehensive, be consistent with program goals, and be validated by program stakeholders (the essence of Standard 2). The first three chapters have laid out a process to answer the first of the two central questions.

The next three chapters discuss the resolution of the second central question—*How can we do better at ensuring that students learn these skills?* Engineering programs need to provide an education that is better at teaching not only disciplinary fundamentals, but also personal and interpersonal skills, and product, process, and system building skills. In almost all cases, we need to do better within the resources allotted. In order to reach these goals, a program retasks the available resources to get more out of them—it retasks the curriculum and the workspaces, and restructures the learning experiences. This chapter will discuss how a CDIO program is built around an integrated curriculum that incorporates an introduction to engineering. Chapter 5 explains how a program also incorporates experiential design-implement exercises, often in a modern engineering workshop. Chapter 6 describes how the CDIO approach incorporates active learning, as well as integrated learning activities that simultaneously teach disciplinary knowledge, personal and interpersonal skills, and product, process, and system building skills. Therefore, as we begin this chapter, we transition from the *what* implied by first central question to the *how* implied by the second.

An integrated curriculum is characterized by a systematic approach to teaching personal and interpersonal skills, and product, process, and system building skills. In general, an integrated curriculum has the following important attributes:

- It is organized around the disciplines. However, the curriculum is retasked so that the disciplines are shown to be more connected and mutually supporting, in contrast to being separate and isolated.
- The personal and interpersonal skills, and product, process, and system building skills are highly interwoven into mutually supporting courses, relieving the potential tension between technical disciplines and these skills.
- Every course or learning experience sets specific learning outcomes in disciplinary knowledge, in personal and interpersonal skills, and in product, process and system building skills, to ensure that students acquire the appropriate foundation for their futures as engineers.

Said another way, the integrated curriculum forms an *educational system* that has an impact greater than the sum of its parts, The educational system is coordinated, with well-understood and mutually supporting elements—each element taking on a well-defined function. All of the elements work together to enable students to reach program learning outcomes. An important part of an integrated curriculum is an introductory course in engineering, which excites students about engineering; teaches some early key skills; creates a set of concrete engineering experiences on which students can base subsequent learning; and suggests the framework of the education to follow. As with any well-defined system, the curriculum must be designed with the appropriate balance of flexibility and efficiency, It would be a grave error to design a curriculum that leads to precise learning outcomes, but leaves students little choice or flexibility.

This chapter describes the curriculum design process developed and implemented in the CDIO Initiative. The process respects pre-existing conditions and available resources that characterize each individual program, but suggests approaches and alternatives to curriculum design that better support the intended student learning. The first part of this chapter underscores the importance of an integrated curriculum, as defined in Standard 3. It is followed by discussions and examples of systemic approaches to curriculum design, The second part of this chapter discusses the task of introducing students to engineering, and gives examples of how to do this in an introductory course, as defined in Standard 4, Design-implement experiences and pedagogical aspects of integrated learning are discussed in Chapters Five and Six, respectively.

CHAPTER OBJECTIVES

This chapter is designed so that you can

- explain the rationale for a curriculum that integrates learning and establishes learning outcomes that require integration of personal and interpersonal skills, and product, process, and system building skills and disciplinary fundamentals
- lay the foundation for curriculum redesign by benchmarking an existing curriculum and recognizing pre-existing conditions that influence curriculum design in your current setting
- describe the process for designing and implementing an integrated curriculum
- describe the purpose and benefits of an introductory course in an engineering curriculum

THE RATIONALE FOR AN INTEGRATED CURRICULUM

A integrated curriculum is characterized by a systematic approach to teaching personal and interpersonal skills, and product, process, and system building skills organized around, and integrated with, engineering disciplinary fundamentals. This integrated approach to curriculum is the focus of Standard 3.

STANDARD 3 – **INTEGRATED CURRICULUM**

A curriculum designed with mutually supporting disciplinary courses, with an explicit plan to integrate personal and interpersonal skills, and product, process, and system building skills.

Disciplinary courses are mutually supporting when they make explicit connections among related content and learning outcomes. An explicit plan identifies ways in which the integration of personal and interpersonal skills, and product, process, and system building skills with multidisciplinary connections are to be made.

Practical reasons

There are both practical and pedagogical reasons for constructing an integrated curriculum, From a practical perspective, we have few options other than to retask the available time and resources. In a traditional engineering curriculum, it is difficult to add more content or time, especially if the intended learning outcomes are beyond the disciplinary core content. The average student's class load is completely committed for every term, and programs are reluctant to extend the undergraduate experience. Instead, a curriculum has to make dual use of time and resources within disciplinary courses already available, capitalizing on the synergy of the simultaneous learning of skills and disciplinary outcomes.

Pedagogical reasons

In addition, there are sound pedagogical reasons for integrating personal and interpersonal skills, and product, process, and system building skills with disciplinary knowledge:

- *Personal and interpersonal skills, and product, process, and system building skills depend on the context in which they are taught.*
  In engineering education, personal and interpersonal skills are often referred to as generic skills. At one level, these skills are generic. For example, lawyers, doctors and engineers all need to communicate and work in teams. However, at a more concrete level, personal and interpersonal skills, and product, process, and system building skills as used by engineers are learned and practiced in specific contexts. For example, oral communication proficiency in a technical field depends on discipline-specific aspects—being able to apply disciplinary concepts; examine problems at different levels of abstraction; make connections; and explain technical issues for different audiences.
- *Students develop deeper working knowledge of engineering fundamentals by learning related skills.*
  Learning personal and interpersonal skills, and product, process, and system building skills in a disciplinary context reinforces students' understanding of disciplinary content. Learning these skills provides a way to apply technical knowledge, and in doing so transforms the disciplinary knowledge from abstract ideas into working understanding. Therefore, disciplinary knowledge; personal and interpersonal skills; and product, process, and system building skills are all mutually supporting. Learning skills in a disciplinary context enables students to develop deeper working knowledge of the engineering fundaments.
- *Faculty can serve as role models of valued learning outcomes.*
  Engineering faculty are role models for students. If the faculty believe that learning personal and interpersonal skills, and product, process, and system building skills is important, they will include them with the disciplinary learning outcomes of their courses. In this way, as they demonstrate the skills, they give students opportunities to develop them through course activities. It is critical that faculty show students that personal and interpersonal skills, and product, process and system building skills are important and legitimate aspects of engineering.

Attributes of the curriculum design

The outcome of the process to be described below is the design of the integrated curriculum, If properly executed, this design will have the following attributes:

- Program learning outcomes systematically flow down to learning outcomes in each of the educational components—courses, modules, and other elements.

- The educational system components describe how they mutually support the learning of disciplinary fundamentals, and detail how they achieve the desired levels personal and interpersonal skills, and product, process and system building skills.
- The curriculum design is an explicit plan that is adopted and owned by the entire program faculty.

This last attribute is vital for the success of the integrated curriculum. Since education is owned by the teaching cadre as a whole, and executed largely in individual components, faculty and leaders of all components must agree with the plan.

Faculty perceptions of generic skills

When planning an integrated curriculum, it is important to recognize that faculty members may have different perceptions of the value, and place of personal and interpersonal skills, and product, process, and system building skills as part of the curriculum. Faculty who view that these skills are of secondary importance and that they should be taught separately from disciplinary content, may be unwilling to integrate them into their courses. According to Barrie [1], for example, faculty perceptions of generic attributes fall into four hierarchical categories: enabling, translating, complementary, and precursory (see Table 4.1). The categories are not mutually exclusive, but hierarchical, that is, a category includes all the lower categories.

The perceived relation between the skills and the disciplinary content will affect the way that faculty think the curriculum should be designed. Using this classification, Standard 3 emphasizes upgrading our view on the learning of skills from the *Not-part-of-the-Curriculum* and *Associated* categories to the *Integrated* categories, focusing on the interaction of skills

TABLE 4.1. FACULTY PERCEPTIONS OF GENERIC SKILLS

Enabling	Generic attributes are integral to disciplinary knowledge, infusing and ***enabling*** scholarly learning and knowledge.	Integrated
Translating	Generic attributes let students use and apply disciplinary knowledge. They interact with disciplinary knowledge through application, potentially changing and ***translating*** knowledge, and are in turn shaped by this disciplinary knowledge. They are closely related to disciplinary learning outcomes.	Integrated
Complementary	They are useful additional skills that ***complement*** disciplinary knowledge. They are part of the syllabus, but separate and secondary to disciplinary knowledge.	Associated
Precursory	They are necessary basic ***precursor*** skills and abilities, and may need remedial teaching of such skills at university.	Not part of the curriculum

and disciplinary knowledge. For example, the teaching of communications as part of a disciplinary topic might be seen as *Translating*, while the teaching of design could well be considered at the *Enabling* level, since it enables and reinforces knowledge. In both cases, they must be genuinely integrated into the curriculum.

Designing an integrated curriculum is difficult when faculty disagree on the purpose and place of generic attributes. At this point, stakeholder input on the importance of these skills can be critical to developing consensus. Faculty may also need the opportunity to discuss the arguments for an integrated curriculum and reflect on these issues over time. These discussions may serve to identify relevant combinations of personal and interpersonal skills, and product, process, and system building skills, and disciplinary content in preparation for the curriculum design process.

FOUNDATIONS FOR CURRICULUM DESIGN

We use an engineering problem-solving process to structure the redesign of the engineering education. The starting point for the curriculum design process may differ considerably among programs. Transforming an existing program into a CDIO program implies that a number of initial conditions need to be considered. The design of an entirely new program does not necessarily involve so many pre-existing conditions. Irrespective of the starting point, the word *design* is used here to describe the creation of new programs and the transformation of existing programs.

The curriculum design process model

Figure 4.1 illustrates a model for the design of an integrated curriculum. The model calls for a translation of the CDIO vision into a formal set of goals that will provide a foundation for curriculum design. This translation is informed by the desired learning outcomes, pre-existing conditions, and curriculum benchmarking. Curriculum design itself is then defined as the projection of these goals onto the courses and associated learning experiences that formally constitute a curriculum.

In the foundational phase, the starting points are the process of setting the expected learning outcomes and the examination of pre-existing conditions. The content of a CDIO program is defined by the learning outcomes that follow from the Syllabus, described in Chapter 3. The pre-existing conditions include factors such as program purpose and length, high-level program design, and underlying structure of the curriculum. These in turn are informed by national standards, university rules, and program tradition.

As a departure point for curriculum design, a benchmarking exercise examines the existing curriculum to see how it compares with the expectations, that is, the learning outcomes of the Syllabus. The scope of the benchmarking

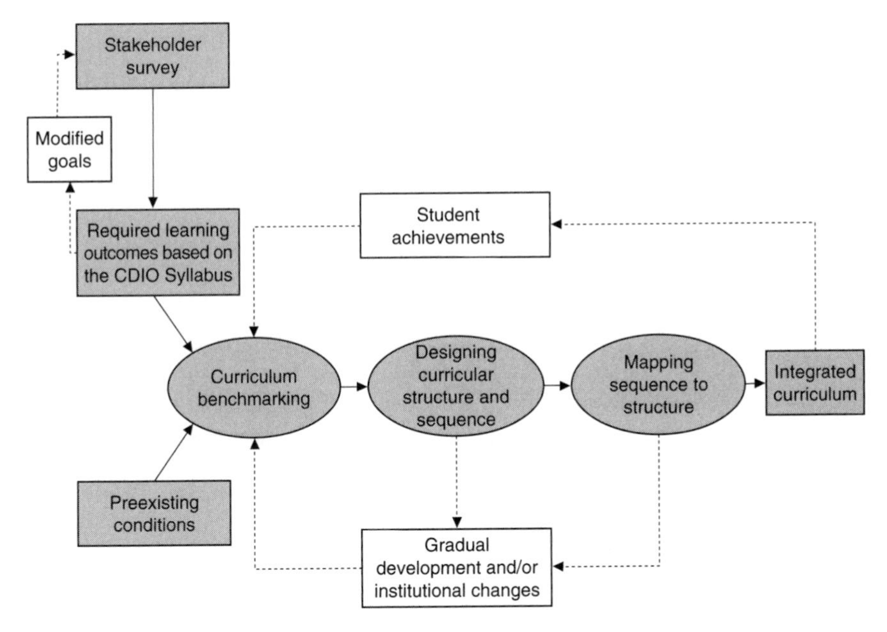

FIGURE 4.1. INTEGRATED CURRICULUM DESIGN PROCESS MODEL

activity includes all of the experiences that contribute to the undergraduate educational experience, For example, institutional humanities requirements may address critical thinking, communication, and ethics. Although outside the engineering program's control, these requirements represent part of a student's education. To some extent, the benchmarking activity can be carried out in parallel with the stakeholder survey of expected proficiencies that was described in the previous chapter.

Once the program goals are clearly established, the pre-existing conditions understood, and the existing curriculum has been benchmarked, curriculum design can truly begin.

Curriculum design proper begins with two parallel, and potentially interacting, steps: the design of the curriculum structure and the determination of the appropriate instructional sequence for each topic. With the structure and sequence established, the last step in design is the mapping of the sequence onto the elements of the structure, so that each element carries well-defined responsibilities for student learning, in an integrated, mutually supporting, and coordinated design. Of course, design is an iterative process with several feedback loops, indicated by the dashed lines in Figure 4.1. Continuous improvement and refinement of the curriculum design is driven by results of student learning assessment, changes in required learning outcomes over time, and institutional changes, such as development funding, altered resources, and faculty re-assignments. In the sections that follow, each of the steps in this design process model is discussed in more detail.

Curriculum content and learning outcomes

The foundation of curriculum design is the delineation of the desired curriculum content and the specification of learning outcomes. Disciplinary curriculum content includes fundamentals of mathematics and the sciences, engineering science, and other technical knowledge, as well as university requirements, such as humanities and social sciences. The Syllabus outlines the content for personal and interpersonal skills, and product, process, and system building skills. In a CDIO program, learning outcomes are specified for disciplinary content, as well as for the skills defined in the Syllabus. Stakeholder surveys identify the expected proficiency of graduating engineers in each of the topics, and potentially for the disciplinary content as well, The process of translating topics into intended learning outcomes is explained in Chapter Three. Learning outcomes and expected proficiency levels, both for disciplinary topics and skills, form the foundation of curriculum design.

Pre-existing conditions

The curriculum design process begins with the need to reflect on pre-existing conditions and the existing connectivity of the curriculum, Pre-existing conditions are the sum of all set factors regarding the current curriculum. These factors include accreditation standards, university rules, program tradition, and requirements of local, regional, and national groups. Three pre-existing conditions strongly influence the amount of flexibility available to the design process: program purpose and length, high-level program design, and the underlying structure of the curriculum.

- *Program purpose and length*
  Programs usually fall into two groups, based on their purpose: those leading to a terminal pre-professional degree for engineering, and those intended to be followed by a subsequent terminal pre-professional degree. This distinction is often reflected in length and structure of the programs. Curriculum design must acknowledge and live within these constraints.
- *High-level program design*
  Institutions often set high-level designs for programs. Some programs are divided into upper-level and lower-level courses. In engineering programs, it is common to find a three-phase or four-phase design: 1) mathematics and science fundamentals; 2) engineering fundamentals; and 3) specialized courses and electives; and 4) summative experiences. (In a three-phase design, summative experiences are considered part of the specializations.) This high-level design affects faculty teaching responsibilities and the degree to which program leaders can influence each of the phases. For example, in many universities, science faculties are responsible for science fundamentals and are not directly influenced by the work of engineering curriculum designers. On the other hand, engineering fundamentals are often taught in required core courses that can be directly influenced. Specializations, or electives, which may or may

not be explicitly recognized as formal curriculum options, are taught by faculty with engineering backgrounds. Because electives are more numerous and scheduled less regularly than core courses, they are harder to influence. Summative experiences, such as final-year projects and design-implement experiences, can provide opportunities to incorporate specializations and electives into the curriculum.

- *Underlying structure of the curriculum*
  Virtually all universities have an underlying structure that dictates the length of the university school year, the length and intensity of the terms or semesters, and the atomic unit of instruction, which is referred to as a *course* in this book. Programs sometimes include units smaller than a course, that is, modules or seminars, or larger than a course, such as a lecture course combined with a laboratory course. Established academic units and total number of allowable units can limit the flexibility of a curriculum plan.

Many of these pre-existing conditions are partially or completely beyond the control of the curriculum designer for an individual program. Curriculum design must address these pre-existing conditions, and the curriculum design process must provide a versatile and flexible approach that will be applicable in the presence of these conditions.

Another form of pre-existing condition is the program's disciplinary content and the degree to which it is fixed. It is important to understand the connections among the topics, that is, existing interactions or isolation of the disciplinary topics within the courses. Written program descriptions and plans for student pathways through the program can provide curricula with mutually supporting disciplinary subjects, as called for in Standard 3. However, the most reliable means of understanding disciplinary structure is to interview program faculty. As an example, Box 4.1 describes the results of a survey of pre-existing conditions in the Applied Physics and Electrical Engineering program at Linköping University. Typically such results show a high degree of connection among the topics. Not surprisingly in the Linköping example, mathematics courses are used by many subsequent courses. The matrix also reveals other key courses, such as *Scientific Computing*, which are used by many courses, This information is important to have available as the design team begins to consider how the curriculum might be restructured.

Benchmarking the existing curriculum

The purpose of benchmarking is to document how an existing curriculum addresses the expectations and proficiency levels of the skills and to serve as an important resource for subsequent design. Engineering programs, in general, already include activities that relate to these topics, but they are often not well designed, well coordinated, or comprehensive, They often do not allocate enough time to include the learning of personal and interpersonal skills, and product, process and system building skills. Benchmarking identifies the

Box 4.1. Connections Between Mandatory Courses In Applied Physics and Electrical Engineering, Linköping University

A survey of faculty members was conducted at Linköping University to document the preexisting disciplinary connections within the Applied Physics and Electrical Engineering program. The purpose of the survey was to investigate and clarify the connections between the mandatory courses, that is, those in the first three years of the program. Hence, the survey covers mathematics, science, and engineering courses. The survey was carried out by asking faculty members responsible for each course to provide a measure of the connections between his/her course and the courses that preceded them in the program. Note that in such surveys, faculty are much more aware of the connection with previous, or supplier, courses than with subsequent, or customer, courses. Each connection was rated on a four-level scale, ranging from *No immediate connection* (white box) to *Strong connection* (dark gray box). A row in the matrix illustrates how the content of a particular course is used in subsequent courses. A column in the matrix shows the extent to which a particular course uses knowledge from previous courses.

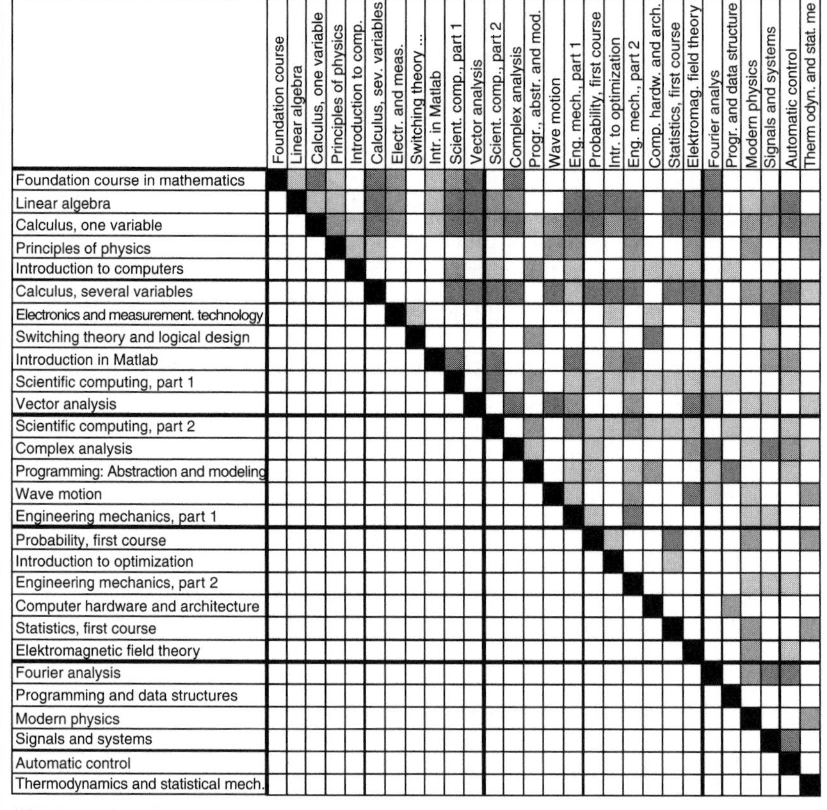

White box: no immediate connection
Dark grey box: strong connection.

– T. Karlsson, Linköping University

TABLE 4.2. **DEFINITIONS OF INTRODUCE, TEACH AND UTILIZE**

	Learning outcomes	Learning activities	Assessment
Introduce	Probably not an explicit outcome	Topic is included in an activity	Not assessed
Teach	Must be an explicit learning outcome	Included in a compulsory activity. Students practice and receive feedback.	Students' performance is assessed. May be graded or ungraded.
Utilize	Can be a related learning outcome	Used to reach other intended outcomes	Used to assess other outcomes

existing committed resources and highlights ways that a curriculum can make better use of time. The CDIO Initiative created a benchmarking tool for this purpose of identifying the existing activities [2].

In benchmarking studies, faculty are asked about the extent to which learning outcomes were addressed in their respective courses. For each of the 14 topics at the second level of the Syllabus in Sections 2, 3, and 4—for example, *2.1 Engineering reasoning and problem solving*—faculty are asked about teaching activities in their courses that address that topic. CDIO Syllabus Section 1 refers to disciplinary content of the program. At the option of the local program, these topics can be included in the benchmarking studies as well.

Teaching activities are categorized as Introduce (I), Teach (T) or Utilize (U), based on intent, time spent, and explicit linkages to learning outcomes, assignments, and assessment criteria. The formal definitions for Introduce, Teach, and Utilize are shown in Table 4.2. The decision to make distinctions among Introduce, Teach, and Utilize was made after observing that the word *teach* is used to describe a great variety of activities occurring within courses.

Face-to-face interviews are conducted with faculty responsible for each course in the program. Responses and data are reduced and analyzed to illuminate patterns of teaching. It is not possible to equate expected proficiency levels of CDIO learning outcomes directly with teaching activity levels. However, it is possible to make some comparisons, to identify weaknesses and strengths in the existing curriculum, and to identify learning outcomes that require more (or perhaps less) emphasis across the curriculum. It is also possible to identify topics that are introduced multiple times without any instructor taking responsibility for actually teaching them. These topics are prime candidates for improvement. Results of benchmarking studies provide important information for curriculum design.

INTEGRATED CURRICULUM DESIGN

Having established the curriculum content and learning outcomes, the key aspects of curriculum design are structure, sequence, and mapping. Curriculum structure refers to the organizational framework of all courses and learning experiences. Sequence suggests the appropriate progression of learning outcomes, and mapping is their assignment to specific courses and learning experiences. Each of these three steps to design—structure, sequence, and mapping—is discussed separately here.

Curriculum structure

Curriculum structure is the arrangement of content and associated learning outcomes into instructional units, or courses, to facilitate intellectual connections among the courses. The requirements for curricular structure in a CDIO program follow from Standard 3. The curriculum structure must allow the disciplinary courses to be mutually supporting, and it must allow the personal and interpersonal skills, and product, process and system skills to be interwoven in the engineering curriculum. A CDIO approach reforms the curriculum to more readily make dual use of time, so that learning experiences enable students to develop both a deeper working knowledge of the fundamentals and the necessary personal and interpersonal skills, and product, process, and system building skills. Several levels of decisions must be made about curriculum structure to support implementation of a CDIO approach. These include choosing the organizing principle, the master plan for integration, the use of block course structures, and a curriculum concept.

Organizing principle. The highest-level choice in integrated curriculum design is that of the organizing principle of the curriculum. Figure 4.2 shows four approaches to curriculum organization. In the figure, disciplines run vertically, and projects and skills run horizontally. A traditional disciplinary organization is depicted at the far left, with the disciplinary topics in isolated "stovepipes." This curriculum organization is the limiting case of an engineering-science approach. Students learn a sequence of topics, with few linkages or interactions, and little integration of skills. In contrast, at the far right of Figure 4.2 is

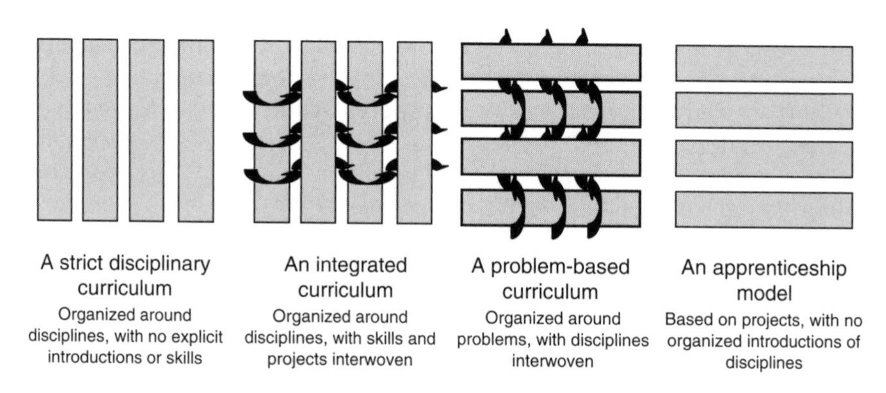

FIGURE 4.2. FOUR APPROACHES TO CURRICULUM ORGANIZATION

a traditional apprenticeship model, in which a student works as an apprentice on a first project, then a second, moving on with little or no formal organization around disciplinary learning.

The middle two options for organization allow integration. The problem-based curriculum uses problems or projects as the organizing principle, integrating disciplinary content on a need-to-know basis, through both formal and informal instruction. Several universities, notably Aalborg University in Denmark, have been quite successful with this curriculum model, and it merits examination. While it is possible to design a curriculum on this model, there are two concerns. The first is that this organizing principle may de-emphasize the technical disciplines, conflicting with the goal of developing deeper working knowledge of the technical fundamentals. The second concern is more practical. Since many universities have a pre-existing disciplinary organization, it may be difficult to transform an existing program to one with a comprehensive problem-based organization.

Therefore, the recommended organizing principle for an integrated curriculum is the model, indicated in Figure 4.2, in which mutually supporting disciplines, with projects and skills interwoven, serve as the organizing principle. This curriculum structure promotes the learning of disciplinary content, and allows several flexible structures for integrating project work and design-implement experiences.

Master plan. All good designs require a master plan of how disciplinary content and learning outcomes will be integrated into the curriculum, Again, several alternatives are possible. Figure 4.3 illustrates a notional school year from left to right in two terms, with the teaching of skills shaded in. The greatest degree of integration occurs in the *integral* model. The learning of personal and interpersonal skills, and product, process and system building skills is totally embedded in the disciplinary courses. All teaching is dual use, strengthening disciplinary knowledge and CDIO skills. The figure depicts an ideal plan.

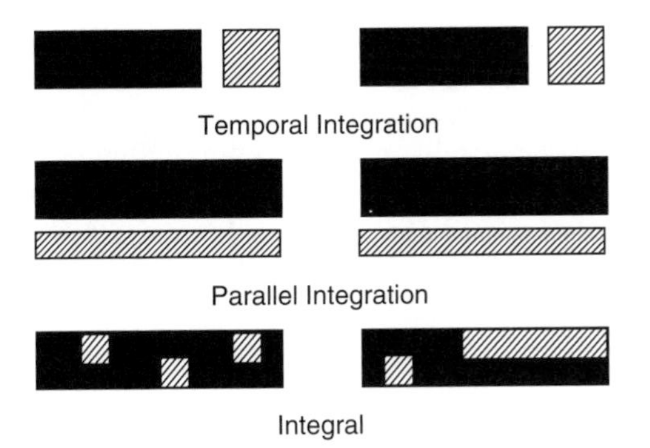

FIGURE 4.3. ALTERNATIVE MASTER PLANS FOR CURRICULUM STRUCTURE

Some use of this plan is important, but it also has limitations in terms of the necessary coordination, especially in the elective curriculum.

A second master plan is called *parallel integration* model. Here a segment of the learning experience in one or more terms is organized around projects or skills, where disciplinary content is taught in parallel with skills. Design-implement courses that span one or several terms would be an example. A third master plan is *temporal integration* model, in which a block of time is set aside for project- or skills-intensive work. Universities that have three terms per academic year or inter-term sessions may find this plan appealing. We have found that some combination of the integral plan plus one of the parallel or temporal integrations best suits our programs, The choice depends largely on local pre-existing conditions. The fact that in the temporal and parallel integration plans the project and skills activities are separate does not in any way imply they do not have dual impact. Well-designed design-implement projects and other experiential learning activities can motivate and reinforce disciplinary learning even if they are in separate modules of the curriculum.

Resources of student time that we do not normally consider in curriculum design are extra-curricular activities during summer terms. There is nothing in Figure 4.3 that suggests that the blocks of time in the temporal and parallel integration schemes could not be extra-curricular projects or summer work. In fact, the time students invest in these activities during their years at university represents a significant share of the time spent in formal educational settings. Since the vast majority of these activities are optional at most universities, the challenge to the curriculum design team is to determine ways in which the extra-curricular and summer time can enrich the learning of both disciplinary knowledge and skills, without making participation a requirement. We recommend fostering appropriate extra-curricular and summer programs, developing resource materials that allow students to more directly understand these linkages, and encouraging students to develop self-learning guides that promote integration of these experiences into their education.

Block course structure. Virtually all universities segment the curriculum into some form of modules or block course structures. Faculty and program managers tend to take this structure for granted because they have limited influence over it. The curriculum is built of instructional units, or courses, of a specified length of time with a specified number of hours. This conventional structure is shown schematically in Figure 4.4(a). Often, the only recognizable connection among courses is determined by prerequisites, that is, courses that must be taken chronologically before others. Sometimes universities allow co-requisites, that is, courses that must be taken before or in the same term as other courses, These are weak temporal structural linkages that do not necessarily reflect any real integration of learning topics in the courses.

The conventional curriculum structure has two major drawbacks to the design of an integrated curriculum. First, it is difficult to create and ensure disciplinary linkages between topics in a conventional curriculum. Secondly, it is sometimes difficult to incorporate the learning of personal and interpersonal

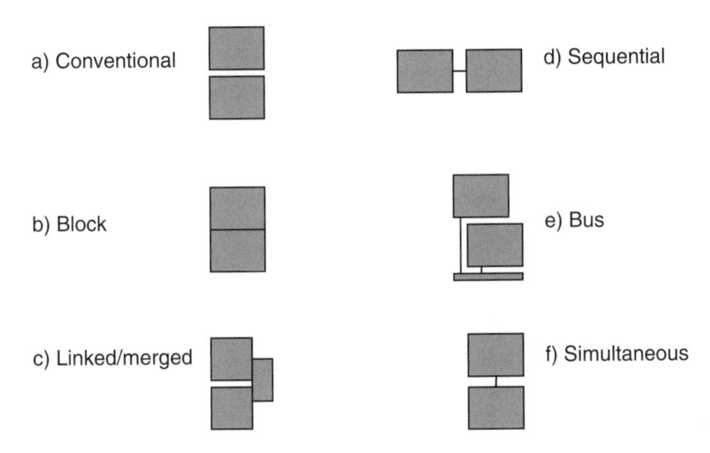

FIGURE 4.4. **ALTERNATIVE BLOCK COURSE STRUCTURES**

skills, and product, process, and system building skills into a conventional curriculum structure. Working within the pre-existing conditions of common university policies and regulations, curriculum designers have identified several approaches to building more flexibility into curriculum structure. Figure 4.4 (b through f) illustrates these alternatives.

- *Block structure (b)*
  Perhaps the strongest of these alternatives is the block structure, in which the time and content allotted to two courses is combined into one course. (Figure 4.4b) Either one faculty member teaches an integrated course, or more commonly, two or more faculty teach together in a closely coordinated fashion. This structure allows a great deal of intra-disciplinary linkages, that is, connections within the course, and tends to make learning experiences across topics more flexible and more common.
- *Linked or merged structures (c)*
  The linked or merged structure allows a disciplinary connection that is almost as strong as the block structure. (Figure 4.4c) In this structure, two faculty members start the term teaching independently, but at some point, the two courses flow together and work in common. This is most effective when the common work is associated with a design project or end-of-term problem that requires the integration of content from both courses.
- *Sequential structure (d)*
  A variant of the merged structure is the sequential structure, in which the time and content allotted to two courses are tightly combined into two consecutive terms. (Figure 4.4d) Here, two faculty members teach as a team, or alternate over the entire length of the two terms, in such a way as to present a more integrated view of the whole. This structure does not allow quite as much flexibility as the block structure, because the allotted time in any given week is that of a single course,

but has the added benefit of fostering the kind of mature understanding of linkages that is facilitated by exposure over a longer time.

- *Bus structure (e)*
  Another structure, shown in Figure 4.4e, is the bus structure. The idea is that some of the allotted time from two or more courses is transferred to a connecting intellectual element that acts as a "bus" for the courses. The bus may be a project, such as a design-implement project, or a set of integrating lectures or seminars. Homework assignments and lectures in conventional courses can be directly related to the bus. One advantage with this design is that students can take the conventional courses without necessarily participating in the "bus" experiences.
- *Simultaneous structure (f)*
  The weakest linkage is found in the simultaneous structure. (Figure 4.4f) In this structure, two faculty members teach two separate, parallel courses. Through good communication and cooperation, they point out, in real time, how learning in one of the courses can influence that in the other. In addition, they individually create exercises that require knowledge and application from both courses.

Except for the conventional structure (a), all of the structures shown Figure 4.4, share the same advantage to varying degrees. They offer curriculum designers flexibility to make disciplinary linkages within the curriculum, and provide opportunities for integrated learning experiences. However, they all share the same disadvantages in that they impose constraints on the flexibility in the path a student takes through the program. Consequently, they are best suited for those parts of the curriculum that are more or less standardized and under the control of the program—the engineering core or a summative design-implement experience. Linkages within a curriculum also place increasing demands on faculty because they require substantial cooperation and adjustments of course content in order to achieve the desired connections.

Concept for curricular structure. Depending on pre-existing conditions and choices of organizing principle, master plan and block structures, a concept for the structure of the integrated curriculum will have evolved. Most likely, an integrated curriculum will contain four types of courses: the introductory course, disciplinary courses, specializations, and summative experiences. An example of a developed concept for curricular structure is illustrated in Figure 4.5. The early part of the curriculum is made up of disciplinary foundational courses and an introductory engineering course that aims to stimulate students' interest in, and strengthen their motivation for, the field of engineering by focusing on the application of relevant core engineering disciplines. In addition, the introductory course also provides an excellent occasion for an early start to the development of personal and interpersonal skills, and product, process, and system building skills. This course accompanies other foundational

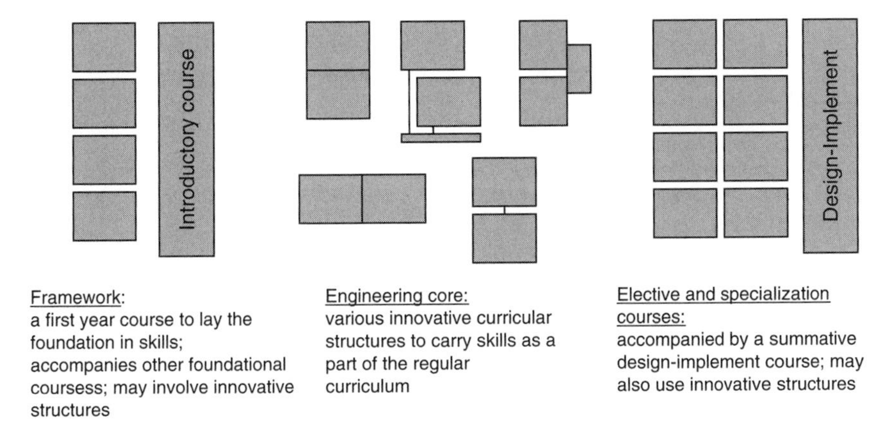

FIGURE 4.5. CONCEPT FOR CURRICULUM STRUCTURE

courses and can involve innovative structures. The introductory course is explained in more detail in a later part of this chapter.

The second part of an integrated curriculum, disciplinary courses, includes engineering courses and related design projects. These often constitute a common or required core of the program. These learning experiences are organized and sequenced into a variety of innovative structures that allow the easier integration of learning outcomes in personal and interpersonal skills, and product, process and system skills together with disciplinary learning outcomes.

The third and fourth parts of the curriculum include specializations, electives, and summative, or capstone, design-implement experiences. As the number of disciplinary courses available to a student expands in this part of the curriculum, integrating skills learning will prove more difficult. In these phases, it is probably best to focus on the summative design-implement course, where a variety of innovative structures can provide flexibility in length of time and sequence of learning experiences. Design-implement experiences are explained in more detail in Chapter Five.

Linköping University follows a similar curriculum structure, as illustrated in Figure 4.6.

Sequence of content and learning outcomes

The next curriculum design issue to consider is the sequence of content and learning outcomes. Sequence is the order in which student learning progresses. If sequence is properly developed, learning follows a pattern in which one experience builds upon and reinforces the previous ones.

In well-established academic disciplines, content sequence is fairly well understood. For the most part, these sequences have been derived from the experiences of faculty who teach and write engineering textbooks. In other

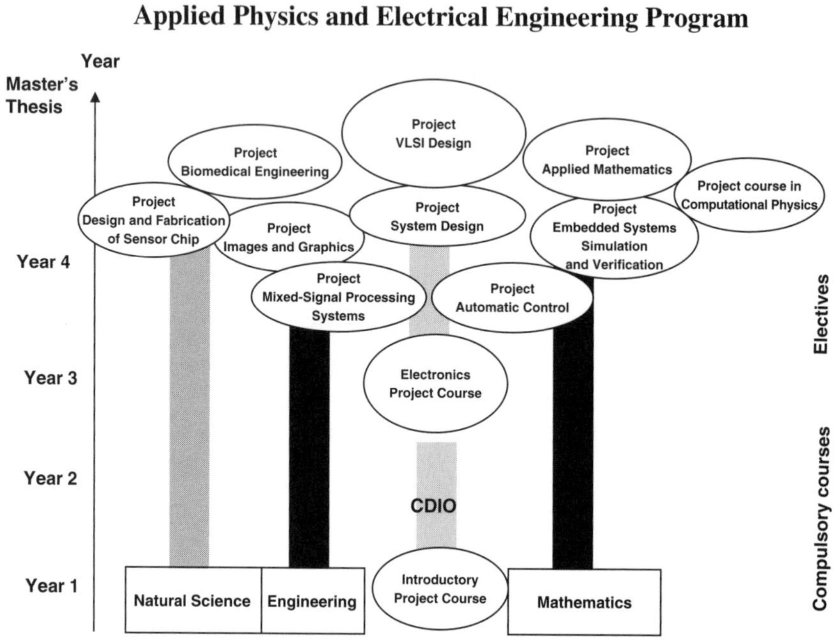

Applied Physics and Electrical Engineering Program

Year
Master's Thesis
Year 4
Year 3
Year 2
Year 1
Project Biomedical Engineering
Project VLSI Design
Project Applied Mathematics
Project course in Computational Physics
Project Design and Fabrication of Sensor Chip
Project Images and Graphics
Project System Design
Project Embedded Systems Simulation and Verification
Project Mixed-Signal Processing Systems
Project Automatic Control
Electronics Project Course
CDIO
Introductory Project Course
Natural Science
Engineering
Mathematics
Electives
Compulsory courses

FIGURE 4.6. CONCEPT FOR CURRICULUM STRUCTURE AT LINKÖPING UNIVERSITY

disciplines, several approaches to sequence exist. For example, some mechanics faculty work comprehensively through 1-dimensional examples before discussing 2- and 3-dimensional examples (also called a strength-of-materials approach). Others introduce equilibrium, compatibility, and constitutive relations, and then specialize to 1-dimension (also called a continuum mechanics approach). The distinctions center on how to proceed on the specialization-generalization spectrum. In newer fields, there is even more variation. For example, in computer science, significant debate takes place about whether it is best to start with the teaching of a programming language, the theory of computation, or the operations of computing machines.

For skills learning outcomes, the appropriate sequence may be less clear. The Syllabus uses a topical organization to suggest what should be taught, Neither the Syllabus, nor the associated learning outcomes, gives guidance on the sequence of topics and skills or the number of repetitions required for proficiency. For example, no sequence is given to teach teamwork. Should the first teamwork exercise be leaderless, with an appointed leader, an elected leader, or a self-selected leader? Should it have a specific deliverable? When should students be taught how to diagnose and negotiate conflict resolution in a team—early or late in their experience? By answering questions such as these, learning sequences for each Syllabus topic can be developed during the curriculum design process. Failure to agree on a sequence at this point complicates the next step, that of mapping the skills onto the curriculum.

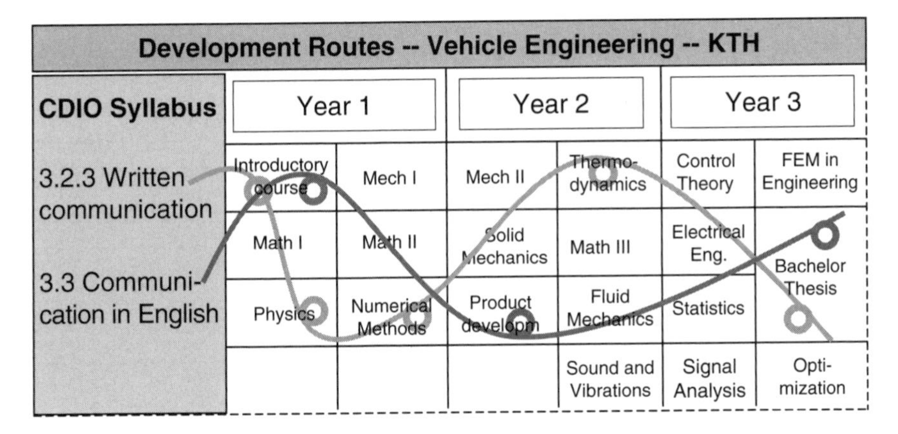

FIGURE 4.7. INTEGRATING SKILLS LEARNING OUTCOMES INTO THE CURRICULUM

A high level of proficiency is often expected of certain complex skills, including design, communication and teamwork. These skills will have to be developed in several courses across the program, For example, in the Vehicle Engineering Program at The Royal Institute of Technology (KTH), several courses in the program integrate teamwork. The sequence of teamwork learning activities is coordinated so that experiences in one course build upon previous experiences and prepare students for the next experiences in the sequence. At KTH, the sequence of learning experiences for a specific learning outcome is called a *development route*. Two of these development routes, for written communication and communication in English, are shown in Figure 4.7.

Coordination of learning outcomes and experiences is necessary both to achieve an appropriate progression of learning, and to use resources and time effectively. For example, while students are learning report writing, they benefit from the repeated use of the same standards for writing technical reports across multiple courses. When they have mastered the standards, they can benefit from increased variation in styles of technical report writing. Joint development between courses gives faculty opportunities to share teaching approaches, feedback forms, and assessment instruments. In addition, with development routes, faculty are more aware of the entire program and the contributions of their respective courses to the whole. Development route "champions" are responsible for helping to develop and distribute materials and for monitoring student progress. Champions can be either subject experts or faculty who have special interest in the respective skills.

Mapping learning outcomes

Once the curriculum structure and learning sequence have been developed, learning outcomes are mapped. Standard 3 calls for "an explicit plan to integrate personal and interpersonal skills, and product, process, and system

TABLE 4.3. AN EXCERPT OF A MATRIX FOR CURRICULUM MAPPING

Course	1	2	3	4	5	6	7
1.1 Knowledge of underlying science							
1.2 Core engineering fundamental knowledge							
1.3 Advanced engineering fundamental knowledge							
2.1 Engineering reasoning and problem solving							
2.2 Experimentation and knowledge discovery							
2.3 System thinking							
2.4 Personal skills and attitudes							
2.5 Professional skills and attitudes							
3.1 Teamwork							
3.2 Communication							

building skills." This explicit plan, or mapping, illustrates how the personal and interpersonal skills, and product, process, and system building skills are woven into the introductory courses, disciplinary courses, and summative courses. Curriculum mapping results in a matrix where one axis lists the Syllabus second-level topics—the X.X level—and the other axis lists the individual courses in the program. Table 4.3 is an excerpt from a generalized matrix of a curriculum mapping. The matrix is filled in with appropriate entries of where each topic will be integrated into the curriculum. Learning sequences and levels of proficiency, which were suggested by stakeholder surveys, determine the appropriate entries.

Faculty involved in teaching the courses in a program must participate in the curriculum design process at this point. They provide insight into the feasibility of integrating specific skills with the disciplinary content for which they are responsible. They also validate the intended sequence of those outcomes. By being part of the curriculum design, throughout its many iterations, faculty develop ownership of the new integrated curriculum.

Guidelines help in deciding which skills to integrate into each course:

- Identify the more natural combinations of disciplinary content and personal and interpersonal skills, and product, process, and system building skills. Some combine more naturally than others.
- Build on the strengths of the existing curriculum.
- Take advantage of where the course is taught in the sequence of the program.
- Start with faculty who are willing and able to develop their courses in this direction. They can set examples and create early successes, which can persuade more reluctant faculty.

Box 4.2 describes an exercise to facilitate coordination between courses. This exercise can be helpful in integrating both disciplinary and skills learning outcomes.

The net result of the integrated curriculum design process should be a curriculum that meets the learning objectives and goals for the program.

Box 4.2. The Black Box: An Exercise to Facilitate Coordination Between Courses

In an integrated curriculum, it is vital that the interfaces between courses be well defined. In this exercise, faculty discuss their disciplines and courses to make explicit the responsibility of each course to students' overall learning. To get to the point quickly, every required course is simply a black box—discussed only in terms of its input and output of knowledge and skills. The point is to keep discussions focused.

Before the meeting, faculty are asked to prepare a short presentation on their respective courses. For each course, they state the specific knowledge and skills students should have as they enter the course, and the specific knowledge and skills students should be able to bring to future courses. These expectations are expressed as intended learning outcomes. With this preparation, it is possible to identify and discuss the connections between courses, adjust inconsistencies, redundancies, and gaps, and reveal places where a course may have drifted from its original intent.

This exercise enhances the dialogue among faculty and highlights the links between mutually supporting disciplines. Development routes for disciplinary and skills learning outcomes are also made visible to all faculty, Benefits of this exercise increase when discussions are well documented. Experience shows that this exercise can spark productive discussions. Therefore, it is wise to allocate an extended block of time at an off-site location to benefit most fully from this exercise.

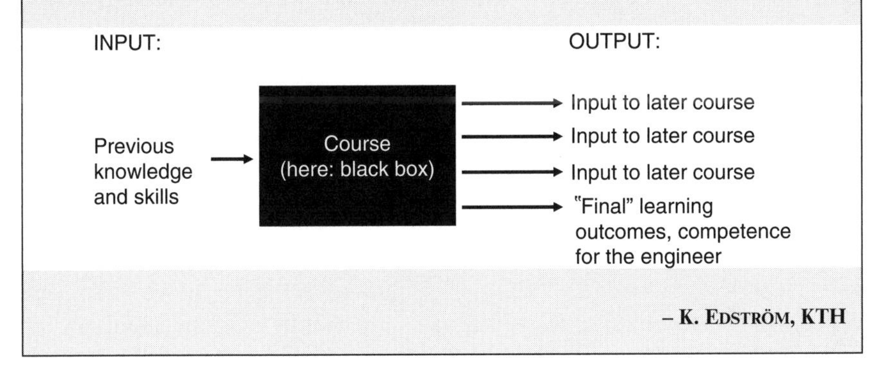

– K. Edström, KTH

Integrated curriculum design requires the three desired features discussed in the introduction to this chapter: mutually supporting disciplinary courses; highly interwoven learning of skills; and well-defined learning outcomes for each course in both skills and disciplinary knowledge. The curriculum design also creates an environment in which integrated learning experiences that make dual use of time can be executed. Integrated learning experiences are discussed in Chapter Six.

INTRODUCTION TO ENGINEERING

As described earlier in this chapter, an integrated curriculum consists of three parts: the introductory course, disciplinary courses, and specializations and summative experiences. The introductory course is an early engineering course that aims to establish the framework in which engineers work and contribute to society. It serves to stimulate students' interest in, and strengthen their motivation for, the field of engineering. Students usually elect engineering programs because they want to create and build. Introductory courses can capitalize on this interest. In addition, introductory courses provide an early start to the development of the personal and interpersonal skills, and product, process, and system building skills described in the CDIO Syllabus. Standard 4 highlights the importance of an introductory course.

STANDARD 4 – **INTRODUCTION TO ENGINEERING**

An introductory course that provides the framework for engineering practice in product, process, and system building, and introduces essential personal and interpersonal skills

The introductory course, usually one of the first required courses in a program, provides a framework for the practice of engineering. This framework is a broad outline of the tasks and responsibilities of an engineer and the use of disciplinary knowledge in executing those tasks, Students engage in the practice of engineering through problem solving and simple design exercises, individually and in teams. The course also includes personal and interpersonal knowledge, skills, and attitudes that are essential at the start of a program to prepare students for more advanced product, process, and system building experiences. For example, students might participate in small team exercises to prepare them for larger product development teams.

The metaphor of building a vault is used to illustrate the role of the introductory course in an integrated curriculum. (See Figure 4.8) The introductory course is the arch-shaped wooden form, or centering, used to support a stone arch made from the disciplinary courses as they are laid in place. When the arch is nearing completion, the capstone, or summative design-implement experience, locks the structure into place. Once the capstone has been added, that is, all disciplinary courses completed, the centering can be removed. Building a real arch without centering is impossible. The introductory course is similar to

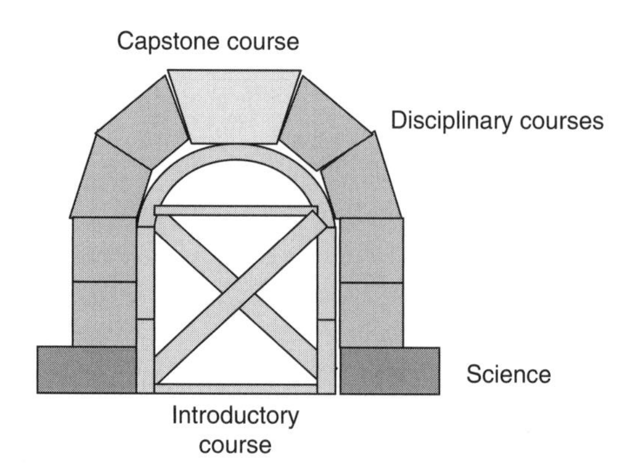

FIGURE 4.8. **METAPHOR OF AN INTEGRATED CURRICULUM STRUCTURE**

centering in that it gives students a quick insight into engineering practice and the roles of engineers. Like the centering, it gives an early idea of the finished shape. The introductory course teaches some essential skills, and provides a set of early authentic personal experiences that motivate the need for disciplinary content, and allow early fundamentals to be more deeply understood. This aspect will be discussed as part of experiential learning in Chapter Six.

Although contained in programs of various engineering disciplines, introductory courses in CDIO programs have a number of common denominators. They are among the first building blocks of their respective curricula. All employ some sort of authentic experience. Some use case studies to discuss historical or contemporary engineering issues. Others use "dissection," that is, taking apart an engineering device, such as a car, to understand how it works. However, most introductory courses include a design-implement experience of some kind that is carried out by student teams of from two to six members. In these cases, design-implement experiences tend to account for at least 50% of the total time devoted to the course. The courses and projects can focus on different stages of product or process development. Some concentrate on one or two stages, for example, design or conceiving and designing. Others address all phases of development—from conceiving through operating. Because it is important for students in these courses to see that their projects actually operate [3], several programs provide student workshops that permit students to build prototypes as part of their introductory courses.

Experience from our introductory courses supports the idea that design-implement projects improve students' comfort level working on technical problems that have no clear solutions. Moreover, students are able to demonstrate an understanding of how to design and build a device from an unidentified assortment of parts [4]. Students welcome opportunities to develop their own ideas in a project, and they appreciate the possibility of seeing something that they themselves have conceived become a reality. Box 4.3 describes an introductory course at Linköping University.

Box 4.3. An Introductory Course at Linköping University

As a result of the CDIO Initiative, the Applied Physics and Electrical Engineering Program at Linköping University created an introductory course called Engineering Project Y. The development of the course started in early 2001, and the course was offered for the first time in 2002. The course has approximately 150 participants each year, runs over the entire first semester, and corresponds to about 25% of a student's workload. This introductory course consists of three main parts: a series of lectures and seminars, project work, and a project conference.

- *Lectures and seminar:*
  The series of lectures and seminars address topics related to the role of an engineer, group dynamics, oral and written communication, information retrieval, and a project management model developed at Linköping University. In addition, representatives of industry give a number of guest lectures.
- *Project work:*
  Project staff assign students to groups of five or six students to carry out the project work. Each group is assigned a project task, defined by a requirement specification. Each year there are approximately ten different project tasks assigned by five different

(*Continued*)

BOX 4.3. AN INTRODUCTORY COURSE AT LINKÖPING UNIVERSITY—CONT'D

departments. Representative projects include: "Web-based supervision of indoor climate," "Detection of moving objects in an image sequence," and "Control system for optimal performance of a model car." Project work is managed by a project management approach, called LIPS, that was developed at Linköping University. The LIPS project model is shown here and explained in detail in Chapter Five.

Starting from the requirement specification, the first step in the project work is to create a project and time plan. These plans must be approved by the project customer before the project work can start. In many cases, group members agree on a group contract that specifies the rules for the work in the group and ways to handle conflicts. During the project work, groups have regular meetings, and they deliver project results to the customer according to the requirement specification. At the final delivery, experts in oral and written communication assess group presentations and give feedback and advice to the groups concerning their communication skills. When the customer approves delivery, each group writes a reflection document in which they evaluate their work, both in terms of the technical result and the group work.

- *Project conference:*
  The course ends with a final conference in which the groups present their work. The conference is organized in parallel sessions with faculty members acting as session chairs. The conference gives students practice in speaking in front of a large audience.

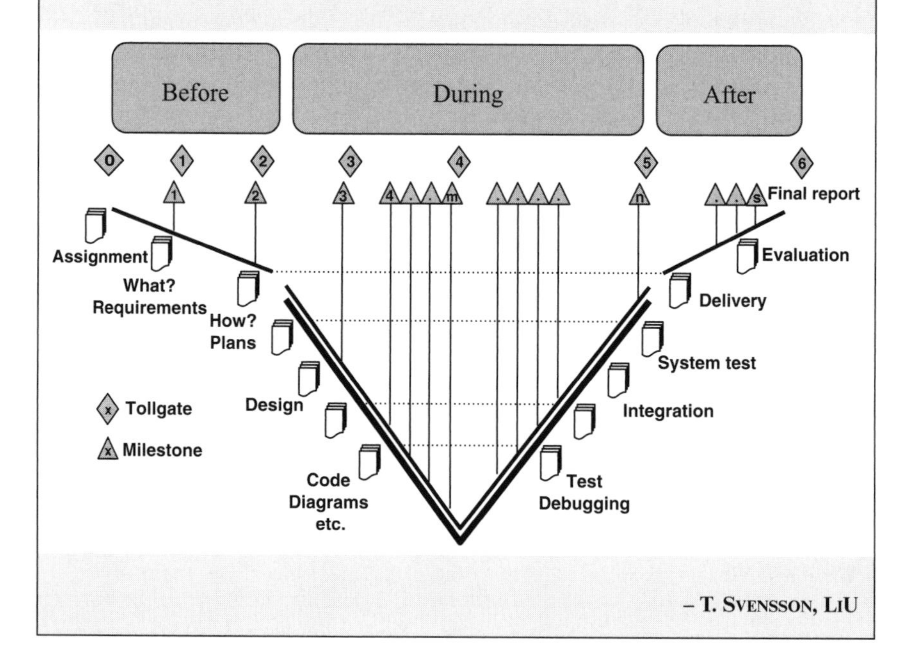

– T. SVENSSON, LIU

SUMMARY

An integrated curriculum is characterized by a systematic approach to teaching personal and interpersonal skills, and product, process, and system building skills. It is organized around, and integrated with, mutually supporting engineering disciplines. This integrated approach promotes deeper working knowledge through application of engineering concepts and emphasizes the importance of skills in engineering practice. Designing an integrated curriculum begins with setting learning outcomes based on stakeholder input, and an examination of pre-existing conditions, such as, program purpose and length and university policies and culture.

The curriculum design process itself focuses on three key components: structure, sequence, and mapping. Examples from CDIO programs illustrate the application of these components. The result of the design process is an integrated curriculum comprised of an introductory course, disciplinary courses, specializations, and summative design-implement experiences highly interwoven with skills learning outcomes. Introductory courses serve to convey the framework of engineering practice, engage and motivate students, teach early skills, and create a set of personal experiences that strengthen disciplinary learning. Introductory courses often include design-implement experiences, which are discussed in more detail in Chapter Five, The issues of designing an integrated curriculum reappear in Chapter Six where the challenges of integrated teaching are addressed.

DISCUSSION QUESTIONS

1. What pre-existing conditions, such as program structure, facilitate or hinder the design of an integrated curriculum in your program?
2. In what ways can you integrate skills learning outcomes into your existing curriculum?
3. Of the many alternative curriculum structures presented in this chapter, which are feasible for your program?
4. How can an existing introductory course be modified to address the purposes of introductory courses in CDIO programs?

References

[1] Barrie, S. C., "A Research-Based Approach to Generic Graduate Attributes Policy", *Higher Education Research & Development*, Vol. 23, No. 3, August 2004.

[2] Bankel, J., Berggren, K. F., Crawley, E. F., Engström, M., El Gaidi, K., Wiklund, I., Östlund, S., and Soderholm, D., "Benchmarking Engineering Curricula With the CDIO Syllabus", *International Journal of Engineering Education*, Vol. 21, No. 1, 2005, pp. 121-133.

[3] West, H., "A Criticism of an Undergraduate Design Curriculum", *Design Theory & Methodology, ASME DE*, Vol. 31, 1991, pp. 7-12.

[4] Newman, D. J., and Amir, A. R., "Innovative First-Year Aerospace Design Course at MIT", *Journal of Engineering Education*, Vol. 90, No. 3, 2001, pp. 375-381.

CHAPTER FIVE DESIGN-IMPLEMENT EXPERIENCES AND ENGINEERING WORKSPACES

WITH P. W. YOUNG AND S. HALLSTRÖM

INTRODUCTION

In this chapter, we continue our discussion of the resolution of the second question central to the improvement of engineering education—*How can we do better at ensuring that students learn these skills?* In Chapter Four, we examined how the curriculum can be restructured and retasked, in order to strengthen the links between the disciplines and weave the necessary skills into the curricular plan. In this chapter, we will examine perhaps the most important device to meet the demands placed on an integrated engineering curriculum—the design-implement experience.

Design-implement experiences allow students to design, implement (build, write, manufacture) and test an actual product, process, or system, or some reasonable surrogate. Such experiences are sometimes called design-build, design-build-test, or design-build-fly. In software, students often design and then write code. Courses based on competitions have aspects of design-build-compete. In contrast with traditional "paper" design courses, the essential feature of such experiences is that students actually build the design and verify its effectiveness.

Design-implement experiences are a key feature of a CDIO program. Their importance is highlighted by the fact that:

- They have dual impact, that is, they teach students personal and interpersonal skills, and product, process, and system design and implementation skills, and at the same time reinforce disciplinary knowledge.
- They strengthen the learning of fundamentals, by being presented multiple times within a curriculum, first to introduce and motivate learning, then to provide opportunities for application.
- They involve both active learning—in which students manipulate, apply and evaluate ideas—and experiential learning—in which students take on roles that simulate professional engineering practice, as will be discussed in Chapter Six.
- They can be motivating and fun, attracting students to engineering and retaining them within the course of study once they have enrolled.

Because of this important role in engineering education, design-implement experiences should not be optional, but should be carefully integrated into the curriculum. Students should be engaged in at least two cycles of design-implement opportunities in order to best support their disciplinary and skills learning.

An important complementary aspect of a CDIO program is that it provides workspaces to facilitate hands-on project-based learning. This need not be new space, created for this purpose, but can be retasked space, previously used for classroom or traditional engineering laboratory exercises.

This chapter discusses the main educational means for planning and conducting design-implement experiences. We have drawn examples from the collective experiences of programs participating in the CDIO Initiative. The workspaces, or learning environments, that enable these design-implement experiences are also discussed. Descriptions include key attributes of effective workspaces and suggestions for modifying existing facilities to accommodate design-implement experiences.

CHAPTER OBJECTIVES

This chapter is designed so that you can

- recognize the importance of design-implement experiences and supporting workspaces in an engineering education
- outline the requirements for design-implement experiences and their appropriate learning spaces
- give examples of design-implement experiences in different educational contexts
- discuss the benefits and challenges of design-implement experiences
- adapt existing facilities and resources to improve design-implement experiences and CDIO workspaces

DESIGN-IMPLEMENT EXPERIENCES

A design-implement experience is a series of events in which learning takes place through the development of a product, process, or system. The key criterion for such an experience is that the object created is designed and implemented to a state at which it is operationally testable by students. In this testable state, students verify that the product, process, or system meets its requirements. Then they identify possible improvements.

The meaning of design-implement experience

We use the term *design-implement experience* to signify a range of engineering activities central to the process of developing new products, processes, and systems. These experiences enable students to live through most or all of the activities in the *Design* and *Implement* stages of the product, process, or system lifecycle model. In fact, these two stages constitute the "core" of the product lifecycle. As we discussed in Chapter Two, *Designing* focuses on creating the design, that is, the plans, drawings, and algorithms that describe what product, process, or system will be implemented. *Implementing* refers to transforming the design into the product, including hardware manufacturing, software coding, testing, verification and validation. As students mature, it may be appropriate to include in design-implement experiences appropriate aspects of conceptual design from the *Conceive* stage. *Conceiving* includes defining customer needs; considering technology, enterprise strategy, and regulations; and developing conceptual, technical, and business plans.

The final lifecycle stage, *Operating*, uses the implemented product, process, or system to deliver the intended value, including maintaining, evolving, and retiring the system. Ideally, students would be exposed to aspects of actual operations as well, but experience has shown that this is difficult in an academic setting for all but the simplest devices. Therefore, for both pedagogical and practical reasons, we focus on conceiving, designing and implementing as the key activities of the product lifecycle about which students should learn.

The product to be designed can be built of hardware, software, or a combination of the two. Media and materials used to build a product need to be carefully chosen, but this does not mean that the product has to be implemented in its final form. Depending on the level of the course, the object of design can be a simple functional model or a complex near-production prototype. If the object to be designed and implemented is a more complex system or process, it may be impossible to actually implement the system or process, but alternatives include implementing an element of it, an analog, a scaled model, or, as a last resort, a digital model. Regardless of the details, the object must meet the basic criterion of being designed, implemented, and verified, so as to provide direct feedback to students on the success of their design.

Role and benefit of design-implement experiences

Design-implement experiences play a key role in an integrated curriculum. For that reason, they are the main opportunity for the dual-impact learning, that is, they are the principal opportunity to construct a single learning event that both teaches skills and reinforces understanding of the fundamentals. From the skills perspective, it is in these design-implement experiences that students develop product, process, and system building skills. In addition, these design-implement experiences are a natural setting in which to teach personal skills, as well as interpersonal skills such as teamwork and communication.

From the perspective of fundamentals, design-implement experiences strengthen a foundation upon which deeper conceptual understanding of disciplinary and multidisciplinary knowledge can be built. Earlier design-implement experiences lay this underpinning by engaging students in problem solving, and motivating the need for analysis. In this way, students are eager to learn a theory when it is presented to them, and they better understand its relevance (and limitations). Later design-implement experiences allow students to apply their theoretical learning, and therefore, such experiences promote both deep understanding and long-term retention. Because of their importance, design-implement experiences are the focus of CDIO Standard 5.

STANDARD 5 – **DESIGN-IMPLEMENT EXPERIENCES**

A curriculum that includes two or more design-implement experiences, including one at a basic level and one at an advanced level.

In curriculum design, we must create a sequence of design-implement experiences from basic to advanced levels in terms of scope and complexity. This iteration reinforces students' understanding of product, process, and system development. More importantly, design-implement experiences should be deliberately structured and sequenced to reinforce learning of the fundamentals. The first design-implement experience should be a concrete experience upon which students can reflect, followed by exposure to theory and abstractions in more formal coursework. The next design-implement experience should allow an application of the previously learned technical knowledge, and the concrete exposure for the next cycle of learning. This cyclical model is the motivation for Standard 5 to require "two or more design-implement experiences." In a simple idealized example, a first year design-implement experience would expose students to a problem that requires a certain key disciplinary theory to solve; a 2nd-year course would present that theory; and a 3rd- or 4th-year design-implement experience would require application of that theory.

Opportunities to conceive, design, implement, and operate may also be included in optional co-curricular activities, for example, in undergraduate research projects, internships, or competitive team projects such as Formula SAE (Society of Automotive Engineering) or the AIAA (American Institute of Aeronautics and Astronautics) Student Design/Build/Fly contest.

In addition to teaching students skills and strengthening the fundamentals, the other main benefits of design-implement experiences are that they

- add realism to the curriculum
- illustrate connections between engineering disciplines
- foster students' creative abilities
- provide engineering successes to strengthen students' self-confidence
- are fun and motivating for students

The emphasis on building products and implementing processes in a real-world context gives students opportunities to make connections between the technical content they are learning and their professional and career interests.

Basic design-implement experiences

As explained in Chapter Four, an integrated curriculum includes design-implement experiences as early as possible. Introductory first-year courses usually include basic design-implement experiences. These early experiences have significant positive effects on first-year students. Students are introduced to structured engineering problem-solving with opportunities to apply fundamental engineering principles. In addition, they learn to work in teams and communicate their progress and results. These early experiences also provide an excellent means to introduce disciplinary content that will be taught in succeeding semesters. Early introduction to disciplinary knowledge is effective in building students' enthusiasm for engineering. In these early courses, students can be creative and have fun with their teammates. Students are also able to gain a deeper understanding of the engineering discipline prior to choosing an area of study.

Basic design-implement experiences have benefits for faculty, as well. From these experiences, faculty become familiar with the personalities, maturity, dependability, and unique skills of students at an early stage in their education. Because they get to know students in a more personal way, faculty are able to recognize individuals, and their learning styles, in ways that are not possible in more traditional classrooms. This enhanced personal contact with individual students facilitates advising, mentoring, and assessment.

Advanced design-implement experiences

In contrast, advanced design-implement experiences are usually planned for 3rd- or 4th-year students. They provide real-world opportunities to analyze, design, build, test, and potentially operate engineering systems that function at higher levels of sophistication than basic design-implement experiences. Where basic experiences have multiple small teams applying a fairly limited breadth of engineering knowledge, advanced project teams are usually larger and require a wider scope of engineering abilities. An advanced project can involve a team up to 12 to 15 students working on a single project over one or more academic terms. Sources of these projects include professional research, cooperative projects with industry, and solutions to real-world problems.

Advanced design-implement experiences are technically challenging in all phases of the project. The work includes design and implementation of student-developed components, as well integration, testing, verification and validation in conjunction with commercially available components or those developed by other students. Technical tasks involved are typically at a level that students will encounter early in their career, including using high performance microprocessors, autonomy and control systems, wireless telemetry, precision machining, lightweight structures, user interfaces and the like. The need for both the technical knowledge and the skills outlined in the CDIO Syllabus increases as the complexity of the task becomes evident to students.

Attributes of design-implement experiences

The development and realization of design-implement experiences can be more complex than traditional course development and teaching. The planning and organization of these experiences require consideration of a number of factors, particularly in selecting relevant projects and managing resources. Moreover, the quality of the learning experience depends on the synergistic combination of appropriate workspaces and adequate faculty support. The detailed nature of a design-implement experience depends largely on the engineering discipline from which it is derived. However, the essential and desirable attributes of effective design-implement experiences are relatively common, and are listed in Table 5.1.

Design-implement experiences throughout the curriculum

We have observed that a single design-implement experience, no matter how well planned, is insufficient to provide students with a complete understanding of the design-implement process. To master these skills, like any other skills, requires practice. An effective strategy is to include a sequence of design-implement experiences in the curriculum and to plan for systematic variation across each instance. Early projects introduce basic concepts and strategies. In later experiences, more complex projects help students integrate knowledge and skills acquired across the entire curriculum. One project might emphasize creativity, while another one stresses manufacturability or multidisciplinary integration issues. Figure 5.1 illustrates a plan that integrates design-implement experiences throughout a five-year curriculum.

TABLE 5.1. ESSENTIAL AND DESIRABLE ATTRIBUTES OF DESIGN-IMPLEMENT EXPERIENCES [1]

Essential Attributes	Design-implement experiences • introduce and reinforce disciplinary knowledge • promote engineering product, process, and system design and implementation skills • focus on design and implementation, including testing • include elements of conception and operation • emphasize learning outcomes rather than the final product or process • allow a number of alternative paths to the solution • are fully integrated into the curriculum • include adequate training in the use of equipment • provide all students with similar opportunities to develop their skills • increase students' motivation for engineering • reward students fairly for their contribution to the task
Desirable Attributes	Design-implement experiences • provide a platform for training in personal and professional skills • develop teamwork skills and build community • develop written, oral, and graphical communication skills • are cross-disciplinary • allow students to build and operate small, medium, and large systems • allow general prototype fabrication, testing, and redesign

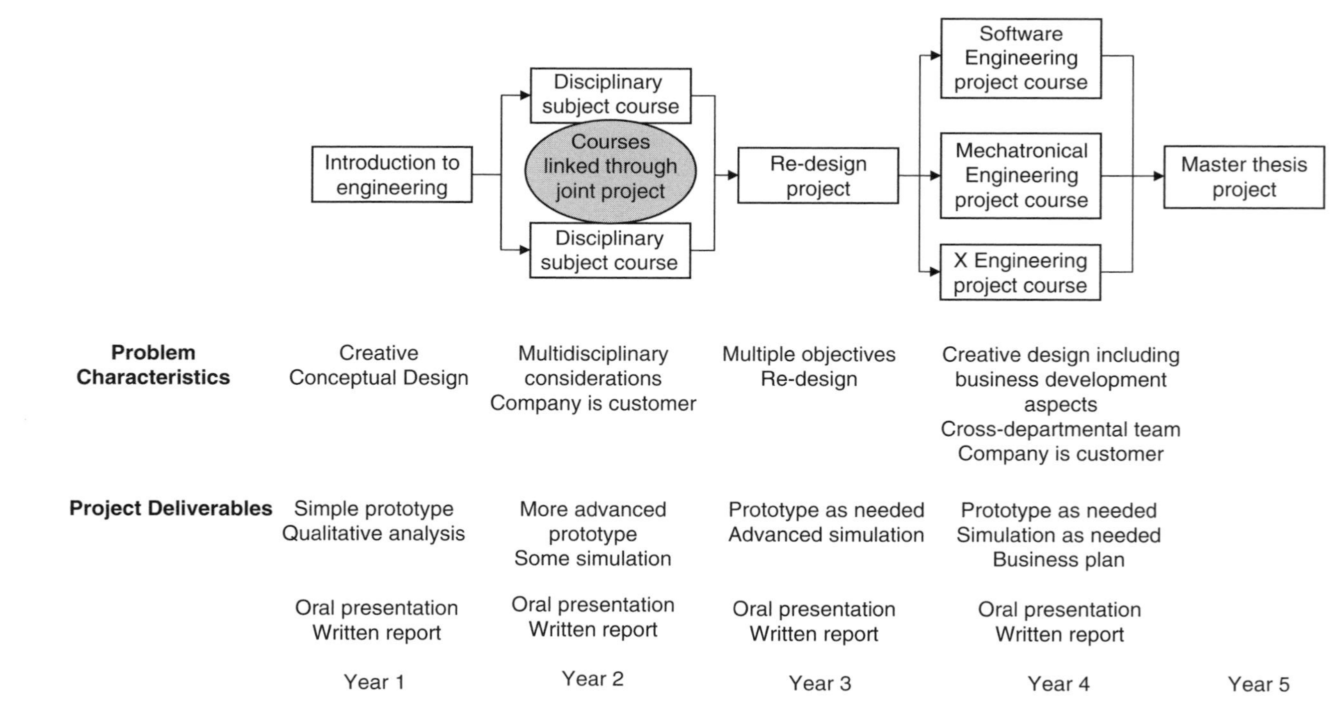

FIGURE 5.1. A PLAN TO INTEGRATE DESIGN-IMPLEMENT EXPERIENCES THROUGHOUT A CURRICULUM

First-year projects. In the first year, the design-implement experience can be part of an introductory course that emphasizes the fundamental principles of the design process, for example, concept generation and selection. Creativity is encouraged through practical exercises. The prototypes are simple, yet enable students to go through the process from user needs to building and testing. The design might include analysis based on fundamentals learned in high school or during the first year. The cost of required materials and equipment are kept to a minimum through proper design of the task. Students typically work in groups of three to five, practicing communication and teamwork skills. A simple project management model can be introduced with specific milestones, nomenclature, and templates for project documentation. An example of such a project model is the Linköping Project Management Model (LIPS) [2], described in Box 5.1.

Second-year projects. In the second year, a design-implement experience can be used to integrate the knowledge acquired in diverse disciplinary courses, as suggested by Figure 5.1. One approach is to pair two engineering subjects, allowing multidisciplinary design. Another option is to pair a disciplinary subject with one that focuses on manufacturability, software engineering or similar topics. Students might design and implement a prototype based on an industrial commission, adding a degree of realism to the task.

At this point, students are able to base their design decisions on the technical knowledge they have gained during their first year of study. They should be able to consider manufacturability of their prototype in order to obtain cost-effective solutions. Simulations can be used to a higher degree than in earlier courses, and prototypes are more advanced. In the second-year experience, communication skills can be explicitly taught, and students again work in small groups. Students present their work orally to the class and share their progress and ideas in written reports. The same project management model, introduced in the first year, can be used again, but on a more detailed level.

In a second-year design-implement project at Queen's University in Belfast, classes are divided into groups of six to undertake the design, construction, and testing of a one-meter beam in three-point bending. The aim is to achieve the highest failure load per unit weight. This experience requires application of theory introduced in the first year and developed in the second year. Box 5.2 is a description of this design-implement experience at Queen's University.

Third-year and fourth-year projects. In the third-year and fourth year, students are given tasks of increased complexity and authenticity. For example, in the third year, they might be asked to redesign existing industrial products in order to improve performance or to decrease environmental load and cost. Analyzing trade-offs among multiple goals is now explicitly considered. At this point, students are able to make decisions using more situation-adapted strategies, selecting prototypes and simulation methods as needed to support the development processes. (See Figure 5.1)

Box 5.1. The Linköping Project Management Model (LIPS)

The Linköping Project Management Model (LIPS), designed at Linköping University in Sweden, supports design-implement experiences, and introduces a professional project management approach in the academic environment. LIPS is aligned with modern industrial project models and is adapted for use in education or small industry projects. The model introduces the phases, definitions, and decision flow needed for running a project in an efficient way. The three phases of the model describe project preparation and planning (the conceive phase), project execution (the design and implement phase), and project delivery and evaluation (the operate phase). The model also includes descriptions of activities, roles, and communication flow in a project.

Electronic templates describe and exemplify project documents. These documents include templates for requirement specifications, project plans, time plans, status reports, meeting minutes, and project reflections. The use of milestones and decision points is introduced. The dynamics in a project are learned through a sponsor-executer relationship. At defined decision points, students are required to deliver documents to get approval to begin the next phase in the project.

LIPS is scalable and can be used to track projects of varying complexity. At Linköping University, the model is used in CDIO courses in the Applied Physics and Electrical Engineering program. Each student uses the project management model for an introductory project during the first year, in a more advanced electronics project during the third year, and in a final project course during the fourth year.

The model has been used successfully with more than 150 projects. Experiences using LIPS yield many benefits. For example, the well-defined steps in the model automatically introduce continuous assessment. They also trigger processes that reveal whether a project is delayed or a group member is not contributing. Moreover, the use of electronic templates greatly reduces the time required to produce and review project documentation.

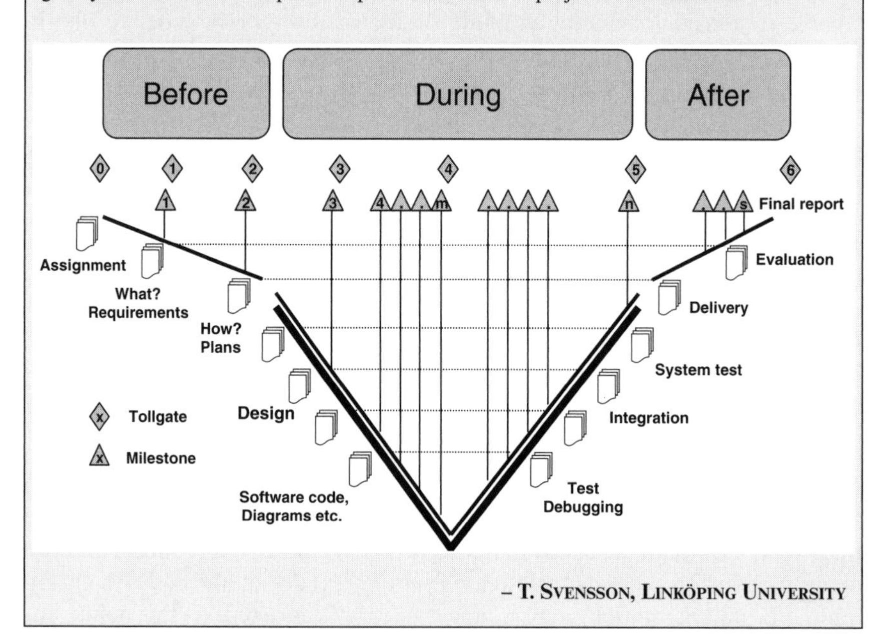

– T. Svensson, Linköping University

In fourth year projects, the scope can be widened further to include business development aspects. Teams consist of larger groups, typically eight to ten students from different engineering departments and possibly students from business programs. Customer and market surveys might be included in the projects. Communication skills and project management models are

Box 5.2. Beam Design Lab At Queen's University Belfast (QUB)

The beam design laboratory at Queen's University Belfast (QUB) is a team-based competition designed to reinforce the fundamentals of basic beam theory. The lab is intended for second-year mechanical engineering students. Design and manufacture take place over a three-week period with a competition at the end of term to find the team with the highest *load at failure-to-mass* design. Each lab period is three hours long, so the complete exercise takes no longer than twelve hours.

Design and Implementation

During the first week, each team of students produces concepts for a beam with specific geometric constraints, without recourse to any calculations. After discussing the concepts with a teaching assistant, the groups choose three designs to build as their prototypes. The prototypes are manufactured from cardboard and tape, and are tested to destruction in three-point bending. On the basis of these results, students choose a final design to be constructed in medium-density fiberboard (MDF).

The second week involves marking out a sheet of MDF in preparation for the cutting and assembly of the final design. While engaged in preparation, student teams must consider the likely mode of failure and calculate the maximum load at which the beam will fail. The third week student teams assemble their beams from the MDF components. The groups have workbenches, some basic hand tools and PVA glue to construct their beams.

Sixteen teams participate in the beam design laboratory. The groups' activities are staggered over a ten-week period to maximize resource utilization. All the teams come together at the end of term for the competition, which now tests the MDF beams in three-point bending until failure.

Student Assessment

Student assessment is a combination of beam performance at the competition (50%), a group report (30%), and a supervisor's evaluation (20%). Beam performance is characterized as the strength/mass ratio of the beams, with maximum credit assigned to the winners, and credit assigned to the other teams on a pro-rata scale. The group report is an ongoing exercise over the term of the laboratory. During the first week, groups sketch their final design and explain the rationale behind it. In the second week, the groups present the calculated mass and predicted load at failure, and explain how they arrived at the figures. The final stage of the report is a reflection, in which groups discuss how the predicted performance of the beam compared with actual performance, and also how the performance of the design could be improved. The supervisor's mark is based on construction quality, design originality, and group dynamics. The final laboratory mark constitutes 10% of the total credit for the Statics course, of which the beam design lab is a part.

The beam design laboratory is a new type of learning activity for students in the first three years and has been judged by students to be quite a success. The lab provides a good mix of teamwork, hands-on experience and applied theory in an authentic setting. From an academic standpoint, not only has it fulfilled its initial goal of reinforcing the fundamentals of beam theory, but has also increased student enthusiasm and improved their perceived relevance of the Statics course overall.

(*Continued*)

Box 5.2. Beam Design Lab At Queen's University Belfast (QUB)—Cont'd

Beam Calculation

Building the Beam

Competition

Post Competition Design Review

– G. Cunningham, Queen's University Belfast

further refined and developed. Organized project management, continuous documentation, and follow-up of decisions are necessary to project completion. Conflicts of interest and multiple approaches to given tasks may be part of the group dynamics.

The project deliverable in the fourth year is an operable prototype or an advanced model, that demonstrates real performance. For example, in a fourth-year design-implement experience at the Massachusetts Institute of Technology (MIT), students designed two autonomous 2 kg robots meant to communicate with each other in space. About 15 students participated in this project intended to complement an existing research program. Students designed, prototyped, and then built or acquired subsystems providing propulsion, navigation, autonomy, and communications. Then they assembled and verified the system. Figure 5.2 illustrates the operational testing of the autonomous robots by students and faculty in microgravity conditions aboard NASA's KC-135A research aircraft. A second example of a fourth-year

FIGURE 5.2. **THE SPHERES PROJECT AT MIT: TESTING AUTONOMOUS ROBOTS IN "0-GRAVITY"**

design-implement experience took place at the Royal Institute of Technology (KTH) in Stockholm, and is described in Box 5.3.

Challenges of design-implement experiences

Providing effective and motivating design-implement experiences for students poses a number of challenges for engineering programs [3].

- *Learning outcomes for design-implement experiences need to distinguish between product performance and learning performance.*
  Learning outcomes, that is, what students will know and be able to do as a result of the design-implement experience, need to distinguish success in acquiring personal and interpersonal skills, and product, process, and system building skills, from successful performance of the product, process, or system that is designed. It is possible to have substantial learning benefits even if the project is not a complete success in terms of a functional product.
- *The task of the design-implement experience must be sufficiently complex, yet limited in scope, to ensure successful outcomes for students.*
  Faculty and students sometimes see the achievement of a good technical solution as the real learning outcome. Failure in the task can be perceived as failure in learning. If the task is too difficult, the result may be an impressive product that is essentially faculty-designed, with students as implementers. A task that is too simple may not motivate students or build the kind of confidence that results from having met a challenge.

Box 5.3. The Project Course at the Royal Institute of Technology (KTH)

The Department of Aeronautical and Vehicle Engineering at the Royal Institute of Technology (KTH) in Stockholm, has developed a design-implement project course for 4th-year students. It is offered jointly for students specializing in Lightweight Structures and Naval Systems. Students are assigned a project task collectively but have the freedom to define and distribute sub-tasks within the project, based on the individual interest and expertise of group members. The course runs over two full semesters. Each student is expected to contribute with technical work in several sub-tasks and to the management of the overall project. A typical group has 10 to 15 students, half from the structures program and half from the naval program. The size of the group inherently creates the need for planning, documentation, information sharing, communication, and team building. The importance of these activities becomes obvious to students during the course of the work.

The project task is carefully designed with respect to Conceiving-Designing-Implementing-Operating:

C The project deliverable should be unconventional, that is, the students should have limited prior experience with similar products. This unfamiliarity enhances the *conceiving* phase and encourages students to inquire into the main technical challenges and approaches. Because the course involves students from different curricular tracks, the project should be multidisciplinary. Different perspectives and incompatible solutions are likely to occur and create the need for compromises and trade-offs among conflicting interests. Consequently, holistic thinking is brought about naturally and in an authentic context.

D The task should be difficult enough to be a true challenge to students, yet be possible to solve if the work is organized and carried out well. One of the main objectives of the course is to encourage students to apply theory and analytic tools and methodologies learned in other courses, thereby solidifying their knowledge and gaining confidence in their roles as engineers.

I The size and complexity of the product should entail teamwork and coordination, and should provide opportunities to obtain practical experience from real manufacturing on a prototype level.

O The task should be formulated in a way that the assessment of the project involves operation of the product and evaluation of its performance with respect to technical specifications. Safety issues are given high priority during operation.

The task is defined by a brief requirement specification of a technical system, typically some kind of vehicle, where the expected performance is described together with constraints in terms of cost, size, weight, power supply, and other factors. The idea is that the requirements are very clearly specified early in the course but the approach to the task is by no means obvious. How the final product will come out is still very open.

A website is used for documentation and information sharing within the project. In this way, all information is continuously updated and immediately shared within the project group. The website also enables interested people outside the project group to monitor the progress of the work. Previous projects have included a human-powered water bike and a subskimmer, capable of submersion and skimming the surface, shown here.

– S. HALLSTRÖM AND J. KUTTENKEULER, KTH

Student time spent on the task needs to be carefully monitored in order to maintain a balance with other competing demands on student time.

- *Design-implement experiences require teaching and assessment practices that are different from traditional instruction.*
  With design-implement experiences, faculty roles change from lecturer and dispenser of information to mentor and coach. In a less constrained learning environment, students are encouraged to discuss, reason, and explore issues with support from faculty. The successful faculty coach is a mentor providing support, a mediator serving as a buffer to clients, and

a manager guiding team and design processes [4]. Methods for teaching and assessing design-implement experiences are addressed in further detail in Chapters Six and Seven.

- *Few faculty are prepared to assume responsibility for technically challenging projects.*
  In a typical engineering department, only a small fraction of faculty and staff have personal, practical experience of developing complex systems. Many programs depend on the talents and skills of one or two key individuals. The introduction of design-implement experiences requires adequate faculty resources to ensure stable and sustainable operation. Some engineering programs hire graduate teaching assistants involved in funded research that has goals and objectives that support the intended design-implement experiences. This approach supplies valuable technical assistance to students, while also accomplishing pre-established graduate research goals. Other programs use technical advisors from industry who are interested in specific student projects. The challenge to enhance faculty competence in skills is addressed again in Chapter Eight.
- *Design-implement experiences need to be cost-effective.*
  Reluctance to include design-implement experiences in engineering education is frequently rooted in suspicions that such experiences are highly resource intensive. Design-implement experiences cost, on average, 1.5 times that of a traditional lecture course, with a span from 1.0 to 2.5 [3]. With some creativity, it is possible to develop lower-cost design-implement experiences without compromising educational outcomes. One university's internal study of the return-on-investment of an advanced-level design-implement-test project concluded that the success of an ambitious project was a key factor in the university's bidding for, and winning of, follow-on research proposals. Their figures support a conclusion that there was a 6-to-1 cost–benefit ratio from this combined linkage of academic and research goals [3].

Stakeholder reactions and summary

Design-implement experiences are one of the key elements of an integrated curriculum. By creating dual impact learning, they support the goals of educating students to master a deeper working knowledge of technical fundamentals and leading in the creation and operation of new products, processes, and systems. They also can be fun, motivating, and the central feature of a student's engineering education. In a study of the impact of design-implement experiences [3], students, faculty, and industry representatives gave these experiences high marks. From an industry perspective, there is evidence that students who have participated in design-implement experiences are positively received because they possess knowledge and skills that are highly valued by industry employers. Table 5.2 lists some representative comments.

TABLE 5.2. REPRESENTATIVE COMMENTS ABOUT DESIGN-IMPLEMENT EXPERIENCES [3]

Students	"I would just like to add that this is the most rewarding course I have ever done as a student!" (KTH)
	"Creating the prototype raised the quality of the work" (Chalmers)
	"It is very good to get experience of running a project in an industrial way"
Faculty	"The students are more motivated" (Chalmers)
	"The (vehicle engineering) project generated an urge for knowledge." (KTH)
Industry	"Excellent students and excellent work! Send more of that caliber!" (SKF-ERC, The Netherlands, about students from Chalmers Mechatronics specialization)
	Fourth-year design-build experience products at MIT reviewed by industrial experts were assessed as being comparable to professional design studies

ENGINEERING WORKSPACES

Providing students with successful design-implement experiences requires a learning environment with adequate spaces, equipment, and tools. We call these facilities workspaces, to suggest their linkage to creative engineering development, and distinguish them from laboratories, traditionally the site of scientific inquiry. Workspaces may be newly built space, or laboratories and rooms retasked from other existing uses. They are multimodal learning environments that support the conceive-design-implement-operate process for simple and complex problems for individual and group projects. They create the infrastructure that visibly signals and supports active and hands-on learning strategies.

Role and benefit of CDIO workspaces

If students are to understand that conceiving—designing—implementing—operating is the context of their education, they need to be immersed in workspaces that are organized around C, D, I and O. We can use the organization of the space to signal the importance of the context to the students, and to strengthen their education. Consequently, workspaces comprise a key element of the CDIO program strategy. Workspaces and other learning environments that support hands-on learning are important resources for developing skills in designing, building, and testing products, processes, and systems. Workspaces are the focus of Standard 6.

STANDARD 6 – ENGINEERING WORKSPACES

Engineering workspaces and laboratories that support and encourage hands-on learning of product, process, and system building, disciplinary knowledge, and social learning

The physical learning environment for a CDIO program includes traditional learning spaces, such as classrooms, lecture halls, and seminar rooms, as well as engineering workspaces. These workspaces aim to support the learning of product, process, and system building skills, while at the same time supporting both disciplinary and multidisciplinary knowledge. They are designed to promote hands-on learning in which students are directly engaged in their own development, and to provide opportunities for social learning. CDIO workspaces are settings in which students can learn from each other and interact with groups. Students who have access to modern engineering tools, software, and laboratories have opportunities to develop the knowledge, skills, and attitudes that support product, process, and system building competencies. These competencies are best developed in workspaces that are student centered, user friendly, accessible, and interactive.

In addition to these direct educational benefits, inviting workspaces that attract students and allow them to work together in stimulating environments strengthen the motivation of students. As the students establish a pattern of working in these spaces, faculty soon start to visit, improving student-faculty interaction. Purely social functions also occur. Thus, workspaces also play a role that goes beyond their initially planned purposes—as community-building spaces for students and faculty.

Designing workspaces

Workspaces are designed to actively engage students in creative and experiential learning and are designed to support the entire curriculum. This is in contrast with conventional student laboratories that, as a rule, are centered on a discipline and/or expect students to take on a more passive role. Traditional student workspaces tend to enable the learning of specific skills, for example *LabView*-supported environments [5], project studios [6]-[7], CAD/CAM/CAE labs connected to workshops [8], and multimedia environments [9], rather than an integrated set of workspaces. Conventional student laboratories tend to be heavily oriented towards demonstrations, tending not to support conceiving, designing or community building.

A CDIO program generally requires new types of workspaces that allow students to work through the entire product, process, or system lifecycle. In this context, *workspaces* denotes facilities that cover a wide span—from traditional student work areas to team-based project areas, to concurrent engineering computer-driven design rooms, to facilities designed for extracurricular engineering activities. The term *workspaces* includes environments that enable students to manufacture mechanical parts, assemble circuit boards, code and load software, and other similar tasks as needed. Workspaces vary significantly between programs and institutions. Therefore, the guidelines given for workspaces identify common criteria, independent of engineering discipline. The guidelines concentrate on the essential attributes and usage modes of workspaces rather than attempt to provide strict blueprints for learning environments. Workspace configuration and size, equipment, and instrumentation depend on available resources. Table 5.3 summarizes the essential and desirable attributes for engineering workspaces in CDIO programs.

If we are to emphasize that conceiving, designing, implementing, and operating is the context of engineering, it would be desirable to make the spaces in which students work explicitly reflect these phases. Facilities need to be flexible and multifunctional, supporting information-based as well as hardware-based projects. Workspaces typically support learning about the four phases of the lifecycle in four different kinds of spaces [10], as illustrated in Figure 5.3.

- *Conceive* workspaces enable students to envision new systems, reach understanding of user needs, and develop concepts. They include both team and personal spaces in order to encourage conceptual development and reflection. Typical equipment and resources include whiteboards, access to online and library resources, and data projectors. Workspaces are largely technology-free zones. Their primary goal is to support the human interactions of talking, listening, and reflection.

TABLE 5.3. **ESSENTIAL AND DESIRABLE ATTRIBUTES OF CDIO WORKSPACES**

Essential Attributes	CDIO workspaces are designed to • encourage hands-on learning of product, process, and system design and implementation, while at the same time supporting disciplinary and interdisciplinary knowledge • facilitate student learning of personal and interpersonal skills • facilitate group activities, social interaction, and communication leading to social learning • comply with local health and safety regulations • provide sustainable resources
Desirable Attributes	CDIO workspaces are designed to • be organized and managed by students • provide flexible equipment, furniture and facilities • facilitate access by students beyond normal class hours • provide access to modern tools, equipment and software

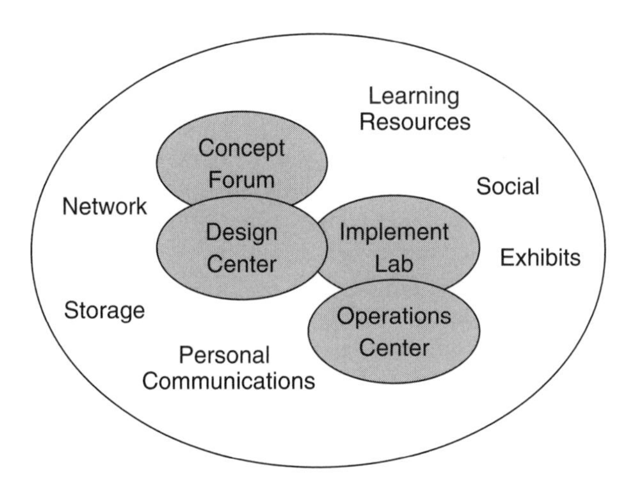

FIGURE 5.3. **CONCEPTUAL MODEL OF CDIO WORKSPACES [10]**

- *Design* workspaces support the new paradigm of collaborative, digitally supported design. They enable students to design, simulate designs, share designs, and understand interactions. Typical equipment includes computers with software for computer-aided design, computer-aided manufacturing, software development and simulation. Additional equipment, for example, videoconferencing and shared databases, may support collaboration with other student teams around the world.. Design workspaces should also be accessible after normal class hours as students often conduct design work during available time at night and on weekends.
- *Implementation* workspaces enable students to build small, medium and large systems, which include mechanical, electronic, and software components. Typical equipment includes metalworking machines, hand tools, measurement equipment, equipment for manufacturing circuit boards, and computers for integrating software into the product. The range of student projects calls for a great deal of flexibility. Safety and accessibility are other critical issues.
- *Operate* spaces create opportunities for students to learn about operations by conducting their experiments and manufacturing their designs. Operation is difficult to teach in an academic setting, but students can learn how to operate both their own experiments and class experiments. Simulations of real operations and electronic links to real operations environments supplement direct student experiences [10].

These physical workspaces should be connected to reinforce their ideological linkage. In addition, they should be connected to other common student facilities, such as the library, social spaces, storage facilities, and through networks to the online community. The workspaces might also include exhibits that reflect engineering research and development projects or that speak to the history of a field or the contributions of a department or academic program.

Examples of CDIO workspaces

CDIO workspaces explicitly do not have to be new, but can use retasked space. Many universities have student laboratories of the conventional kind that are highly underutilized. Retasking some of this space as a CDIO workspace often results in much higher utilization. Based on our experience, as the students engage in active work in the new workspaces, both the attractiveness and need for older conventional passive student laboratories diminishes. It may also be possible to retask classroom and meeting room space into workspaces for projects that do not require large manufacturing equipment.

Implementation of CDIO workspaces at the existing collaborators has ranged from the design and construction of new space to the adaptation and redesign of existing physical layouts. Chalmers University of Technology in Göteborg retasked existing spaces to create a Prototype Laboratory. In this facility, students from programs in Mechanical, Automation and Mechatronics, and Industrial Design Engineering create computer-assisted design models to manufacture prototypes and test functional models of various mechanical and mechatronic products. Prototypes can be made of wood, metal, plastic, cardboard, electronics, and/or software as best suits individual project needs.

Another example of a CDIO workspace is the *Poolen* at the Royal Institute of Technology (KTH) in Stockholm. Converted from previously underutilized class/meeting room, this facility measures approximately 60 m^2 and serves as a combination of design space, meeting room, manufacturing room, and assembly and test room for students enrolled in the Vehicle Engineering specialization. Their most recent project was a water vehicle capable of carrying a human operator at high speed over a water surface and, with minimal changes required, also capable of diving below the water surface to operate as a submarine for extended time periods [11]. (The subskimmer is illustrated in Box 5.3)

A third example of more significant retasking of space, combined with some new construction, is the MIT Complex Systems Laboratory. The Department of Aeronautics and Astronautics established a full suite of workspace facilities to address each specific Conceive, Design, Implement, and Operate element of the CDIO lifecycle model. The existing basement and first floor of the building were renovated, creating a new learning laboratory complex. Specific CDIO functions were designated for each work area: conceiving in the Seamans Concept and Management Forum; designing in the Design Center; implementing in the Gelb Laboratory (containing machine and electronics shops, and a rapid prototyping facility); and operating in the Neumann Laboratory and the Hangar and flight operations center. In the Seamans Laboratory, located on the first floor, there is an additional open area of approximately 1500 m^2 designated for study and for community building. Students study in groups, interact informally with faculty and teaching assistants, and have access to computers to assist them with their assignments. The departmental library is immediately adjacent to this space.

Several key architectural themes are incorporated into these workspace facilities. Movable furniture allows convenient space reconfiguration to meet changing demands of class size, teaching style, and project needs. Electronic door controls give students access to the facilities (other than machine shops) at night and on weekends.

Using the lessons learned from these and other world-class student workspaces, including the Integrated Teaching and Learning Laboratory at the University of Colorado, and the Integrated Learning Center at Queen's University, Canada, other CDIO collaborating institutions have succeeded in developing their own workspaces that support their educational goals. The breadth and scope of each school's workspace facilities vary according to available space, funding, program needs and other factors, but the common theme is the awareness that Conceive, Design, Implement, and Operate workspaces are effective facilitators of improved engineering education.

Teaching and learning modes in CDIO workspaces

Teaching and learning modes in CDIO workspaces fall into three major categories: product, process, and system design and implementation, reinforcement of disciplinary knowledge, and knowledge discovery. In addition, workspaces play a major role in building community among students. For each category, a number of more detailed teaching and learning modes can be described. These modes are meant to be nearly exhaustive, and may be overlapping. They are intended to serve as a guide in thinking through the requirements for workspace design at a specific university. Figure 5.4 illustrates the teaching and learning modes facilitated by these workspaces.

Product, process, and system design and implementation. This category represents the most obvious major mode of teaching and learning that takes place in a CDIO workspace. However, it should also be recognized that there are many variations of this mode with different requirements for the design of the workspace.

- *Basic design-implement projects* are course-based design projects carried out over the period of a semester by student teams from a given course.

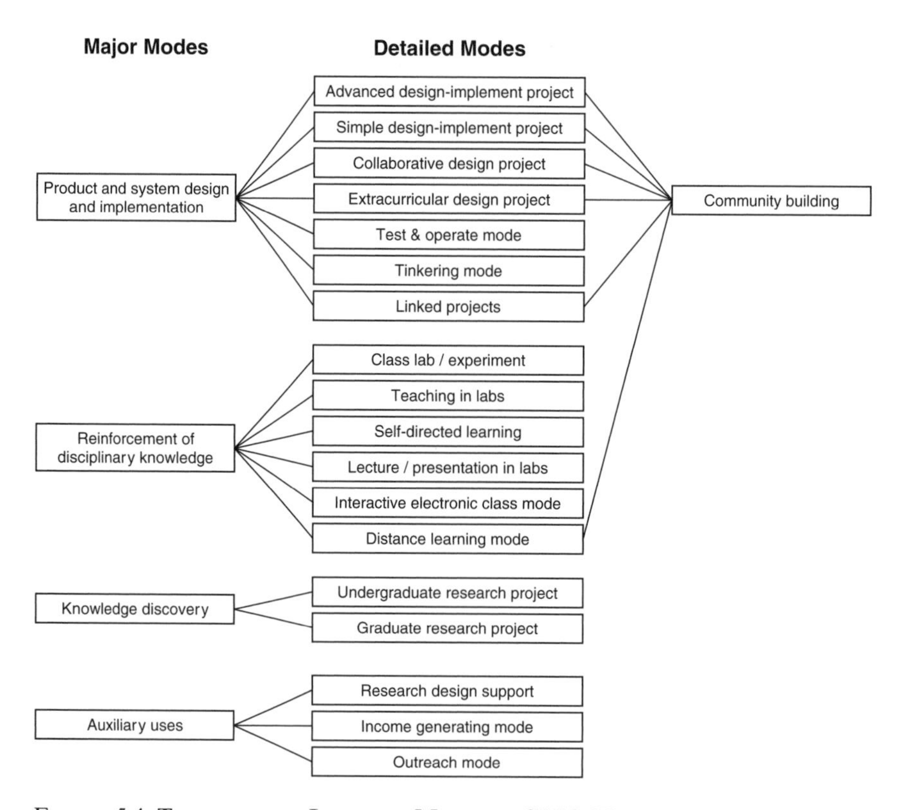

FIGURE 5.4. TEACHING AND LEARNING MODES IN CDIO WORKSPACES

Design work includes computer simulation and visualization. The work results in both a "paper" design and a simple prototype. Typically the work is conducted in smaller teams of 3 to 8 students. Support for this mode includes design tools, management tools, visualization tools, and basic prototyping facilities.

- *Advanced design-implement projects* are design intensive and team oriented, requiring dedicated space for periods of time ranging from a semester to a year. Advanced design-implement projects involve several disciplines and result in a product or prototype consisting of varying amounts of hardware and software. These projects typically take place in the 4th year of the curriculum and are usually conducted by teams of 10 to 20 students.
- *Collaborative design projects* are projects conducted in collaboration with other universities, government, or industry. Projects may be a response to a collaborator's needs, or a partnership in which team members are all working on various segments of the same system. This mode is communication intensive, and requires real-time data, voice, and video communication.
- *Extracurricular design projects* are typically aimed at building something for competition, such as human-powered aircraft, robotic helicopters, or solar cars. Teams come from several engineering disciplines and require office space, design space, building and testing space, storage space, and access to the facilities after hours. These year-long to multi-year projects typically involve teams of 5 to 10 people and result in operational prototypes of significant size.
- *Test and operate* modes are intended to teach students about the operational concepts of engineering systems by giving them hands-on experience in testing and operation. This mode requires personnel dedicated to the maintenance and operation of the systems, long-term dedicated space for the equipment, and real-time communications with other departments and other sites.
- The *tinkering* mode is for individuals working on projects in their spare time. These projects typically require the use of shop equipment, tools, and work surfaces, and happen any time workspaces are open.
- *Linked projects* are longer-term interdisciplinary projects between several sections of the department and/or the university. The projects can connect several courses within a program. For example, an autonomous ground vehicle project may require mechanical prototyping and fabrication using metal working machines, computer-aided design software for mechanical layouts, computer tools for software generation, and hardware testing. Linked projects can also connect different programs, involving teams of students in different specializations working together and contributing their expertise from their respective disciplines. The interdisciplinary nature of this mode requires spaces for meetings, work, storage, and formal presentations.

The variety of modes strongly suggests the need for careful consideration of the links between the curriculum and workspace design and the need for both short-term and long-term flexibility.

Reinforcement of disciplinary knowledge. CDIO workspaces are designed to strengthen students' disciplinary knowledge by providing support for hands-on, active learning strategies that engage students directly in thinking and problem solving activities. (Active and experiential learning are described in more detail in Chapter Six.) There are a number of teaching and learning modes in these workspaces that lead to the reinforcement of disciplinary knowledge.

- The *class lab experiment* is the traditional lab assigned by faculty in which students collect and reduce data. They then write reports on the procedure. These labs are usually conducted in teams of 3 to 6 students and require benchtop set-ups or larger fixed installations.
- The *teaching-in-labs* mode is a demonstration of principles and phenomena with equipment that is specific to the lab. Faculty usually bring the entire class to the lab to give the demonstration. An extension of this mode is the use of electronic classrooms to maximize space and use equipment dedicated to the room, such as networked projection units.
- In the *self-directed learning* mode, students learn engineering concepts and principles on their own. Traditionally, this has meant reading textbooks and doing research in the library. Self-learning now includes educational videos, online information and programs, and other electronic media.
- *Lecture or presentation* is the standard classroom-teaching mode. Faculty use electronic presentation hardware and software in the classroom for showing course material and simulations that demonstrate theories and principles.
- *Interactive electronic class* is a fully electronic classroom where students are able to do computer-based work in real time, with faculty supervision and assistance. Interactive software is used to comment on work, and projection equipment is used to demonstrate examples. Interactive design classes are an extension of this mode.
- *Distance learning* includes videoconference classrooms and broadcast studios in which instruction is delivered to multiple remote sites in real time.

Knowledge discovery. CDIO workspaces can also support student research projects. They do so by making accessible to students a range of equipment usually found only in research labs.

- *Undergraduate research projects* focus on 3rd- and 4th-year student research projects that involve students designing, building, operating, and reporting on experiments with the guidance of faculty advisors. This mode is typically conducted over several semesters, where the first semester is dedicated to background research, and the second to building the apparatus, running the experiment, and reporting results in a formal presentation and document.

- *Graduate research* mode is intended to support a graduate student who needs to establish an experimental set-up for some period of time. The time scale ranges from one semester to several years, requiring that the space be dedicated for that amount of time.

Community building. CDIO workspaces also play a central role in building community among students. In addition to working on their design-implement projects, students use the workspaces to study for disciplinary courses and for informal social functions. CDIO workspaces also provide facilities for student clubs devoted to tinkering, model building, and other extracurricular projects. This is more of an emergent mode that occurs when the previous three modes of use have drawn the students to the workspace, engaged them, and allowed them to interact.

Auxiliary uses. In addition to the teaching and learning modes directly linked to program learning outcomes, there are a number of other auxiliary uses for CDIO workspaces.

- *Research design support* mode. Research teams use the design center capabilities of the workspaces for a short time, that is, hours or days, to work through a segment of the research design. This short-term dedication of space supports the research team's design efforts with analysis tools, design tools, communications, and presentation equipment. This mode may be directly supported by the distance learning mode, where communications equipment is used for meetings with research sponsors.
- The *income-generating external* mode supports external companies who lease the use of specialized experimental facilities. This mode typically lasts for several weeks and requires the dedicated use of the equipment, support staff to operate the equipment, and space. Security of information may be an issue with some companies.
- The *outreach* mode supports public awareness of the CDIO programs and workspaces. Tours of the university include workspaces and explanations of programs that use these facilities. Students might host tours of these workspaces for industry representatives and other guests, explaining ways in which the learning environment supports the program. This mode allows students to present their work, reinforces their learning, and facilitates interactions with industry.

Challenges of engineering workspaces and stakeholder reactions

Integrated learning spaces can provide significant resources, and innovative mechanisms to support the education of engineering students. However they can pose challenges in development and operation. Workspaces can vary significantly with regard to costs and formats, depending on goals, number of students and available financial resources [12]. However, some design and operational challenges stand out regardless of scope. Those challenges are summarized below.

- *The need for a workspace design driven by curriculum and usage modes*
  Young, *et al.* [12] discuss the various usage modes of the workspaces, as presented above. They indicate the multi-purpose nature of engineering workspaces and the central role that they can play in the curriculum. The workspaces can also enable leaders to emerge from among the students in the domains of design, implementation, and experimentation. These students can be involved in the operations of the workspaces—as tutors and as "lead users" to inspire workspace development.
- *Planning for flexibility in usage modes and for enabling the workspace to evolve over time*
  Workspaces need to be flexible to suit the needs of different projects and to facilitate upgrading equipment based on operational experiences. Perceived limitations of the studied workspaces typically involve available floor space or storage space rather than missing equipment.
- *Safety concerns and expanded access for students*
  It is highly desirable to allow after-hours access by students with entries controlled by internal security measures such as ID card controls and keys. While usage of clearly hazardous equipment would be kept to normal academic working hours under staff supervision, engineering workspaces should be deliberately designed to provide an environment for group study, socialization, and mixing of students and faculty in both curricular and extra-curricular settings.
- *Operational scheduling and staffing of the workspace*
  Challenging issues have been identified in the operations of CDIO workspaces. As the attractiveness of the workspace becomes understood and the demand grows, scheduling during the academic semesters becomes a challenge, particularly during highly congested periods of work that emerge at mid-term and end-of-term periods. Acquiring technical staff proficient in a wide number of professional areas, as well as willing to work closely with numerous students throughout the year, is a key ingredient for successful acceptance by the students. Close coordination with academic instructors to plan upcoming workspace projects, as well as to manage ongoing projects, requires diligent effort from all parties.

Student surveys show that students respond positively to workshop environments where they have opportunities to conceive, design, implement, and operate engineering products, processes, and systems as part of the curriculum and in extracurricular activities. Students' response to these workspace initiatives has been uniformly positive at all participating universities.

Survey data taken at MIT, in particular, shows that graduating students in the aerospace program feel that the redesigned workspaces have not only increased their ability to learn disciplinary material, but also have increased their positive feelings towards their classmates and their chosen profession.

SUMMARY

A design-implement experience is a learning event in which learning takes place through the creation of a product, process, or system. These learning experiences play a central role in a CDIO approach. In addition to teaching students how to design, build, and test products, processes, and systems, they add realism to engineering education. Students find that design-implement experiences are fun and motivating. They foster students' creative abilities and strengthen their self-confidence. From a learning perspective, design-implement experiences stimulate learning of technical knowledge, connect theory to practice, illustrate connections between subjects, and enhance the understanding of engineering science. Design-implement projects also serve as vehicles for teaching personal and interpersonal skills, as outlined in the CDIO Syllabus. These educational experiences are highly rated by students, faculty, and industry stakeholders. However, design-implement tasks need to be carefully planned as separate learning events in themselves and also as parts of a planned sequence of design-implement experiences in an integrated curriculum.

CDIO workspaces significantly enhance the education of engineering students. Students respond positively to workspace environments where they experience the four stages of the product or process lifecycle—conceiving, designing, implementing, and operating. These spaces facilitate activities that encourage the learning of design and implementation skills, reinforcement of disciplinary knowledge, and discovery and experimentation. These spaces can vary significantly with regard to costs and format, depending on goals, number of students and available financial resources. However, some design issues stand out regardless of scope: the need for a workspace design driven by curriculum and usage modes, planning for flexibility in usage modes and evolution over time, safety concerns and extended access, and operational issues. The benefits include enabling new approaches to engineering education, the strengthening of student motivation, and improved student-faculty interaction. CDIO workspaces have been shown to play important roles in building social and learning communities that go far beyond their initially planned purposes.

Design-implement experiences, supported by engineering workspaces, are a key part of the integrated curriculum that was presented in Chapter Four. They also support both active and experiential learning, the subject of the next chapter.

DISCUSSION QUESTIONS

1. What design-implement experiences do you offer in your current programs?
2. In what ways would you modify current design-implement experiences or create new ones in light of the ideas expressed in this chapter?
3. How would you address the key challenges to creating and implementing effective design-implement experiences?
4. How can your existing learning facilities and workspaces be modified to support a CDIO approach to design-implement experiences?
5. What specific functions would new workspaces serve?
6. How would you address the key challenges to creating and building new workspaces and learning facilities?

References

[1] Andersson, S. B., Malmqvist, J., Knutson Wedel, M., Brodeur, D. B., "A Systematic Approach to the Design and Implementation of Design-Build-Test Project Courses". Proceedings of ICED 05, Melbourne, Australia, 2005.

[2] Svensson, T., and Krysander, C., "The LIPS Project Model", Technical Report, Linköping University, Sweden, 2004.

[3] Malmqvist, J., Young, P. W., Hallström, S., Svensson, T., "Lessons Learned From Design-Implement-Test-Based Project Courses", International Design Conference—Design 2004, Dubrovnik, May 18—21, 2004.

[4] Taylor, D. G., Magleby, S. P., Todd, R. H., Parkinson, A. R., "Training Faculty to Coach Capstone Design Teams", *International Journal of Engineering Education*, Vol. 17, No. 4 and 5, 2001, pp. 353-358.

[5] Ertugrul, N., "Towards Virtual Laboratories: A Survey of *LabView*-based Teaching/Learning Tools and Future Trends", *International Journal of Engineering Education*, Vol. 16, No 3: 171-180, 2000.

[6] Kuhn, S., "Learning from the Architecture Studio: Implications for Project-Based Pedagogy". *International Journal of Engineering Education,* Vol. 17, No. 4 and 5, pp. 349-352, 2001.

[7] Thompson, B. E., "Studio Pedagogy for Engineering Design". *International Journal of Engineering Education*, Vol. 18, No. 1, pp. 39-49, 2002.

[8] Dutta, D., Geister, D. E., Tryggvason, G., "Introducing Hands-on Experiences in Design and Manufacturing Education", *International Journal of Engineering Education*, Vol. 20, No. 5, pp. 754-763, 2004.

[9] McCarthy, M., Seidel, R., Tedford, D., "Developments in Projects and Multimedia-based Learning in Manufacturing Systems Engineering". *International Journal of Engineering Education*, Vol. 20, No. 4, pp. 56-542, 2004.

[10] Crawley, E. F., Hallam, C.R.A., Imrich, S., "*Engineering the Engineering Learning Environment*", Department of Aeronautics and Astronautics, Massachusetts Institute of Technology, USA, 2002.

[11] Hallström, S., Kuttenkeuler, J., "Experiences From a Three-Semester Design-Implement Project Course", The Royal Institute of Technology (KTH), Stockholm, Sweden, 2004.

[12] Young, P. W., Malmqvist, J., Hallström, S., Kuttenkeuler, J., Svensson, T., Cunningham, G. C., "Design and Development of CDIO Student Workspaces—Lessons Learned", *Proceeding of the 2005 Annual Conference and Exhibition of the American Society of Engineering Education*, Portland, Oregon, 2005.

CHAPTER SIX
TEACHING AND LEARNING

WITH K. EDSTRÖM, D. SODERHOLM, AND M. KNUTSON WEDEL

INTRODUCTION

This chapter broadens and concludes the discussion of the second question central to the reform of engineering education: *How can we do better at ensuring that students learn these skills?* The curriculum design process, presented in Chapter Four, develops an approach to integrating learning outcomes into the curriculum. Design-implement projects, discussed in Chapter Five, are a principal mechanism to create dual-impact learning experiences, and therefore, both fulfill the skills learning outcomes and deepen students' understanding of disciplinary knowledge.

In this chapter, we explore a wider repertoire of teaching and learning methods that are effective in integrating skills with disciplinary knowledge. We start by describing the engineering students' learning experiences, as seen from their perspectives. Then we describe how the skills learning outcomes can be realized through teaching and learning activities. In this chapter, we emphasize the alignment of teaching-learning approaches with curriculum.

Integrated learning means that students practice and learn personal and interpersonal skills, and product, process and system building skills, simultaneously with disciplinary knowledge. While Chapter Four emphasizes a systematic plan to integrate skills into an integrated curriculum; integrated learning focuses on the implementation of that plan in each of the program's courses and co-curricular activities. Design-implement experiences are good examples of integrated learning, but integrated learning is not limited to project-based courses. Integrated learning is an example of active and experiential learning methods that can be applied in a wide variety of disciplinary settings.

Integrated learning experiences and active and experiential learning are fundamental to reaching the educational goals of a CDIO program. The key attributes of these approaches are that

- Planning for integrated learning requires clear specification of intended outcomes related to personal and interpersonal skills, and product, process, and system building skills, as well as disciplinary content.

- Integrated learning places the engineering teacher at the center of student learning of both the technical discipline and skills, and emphasizes the value and linkages of both parts of the education.
- Experiential learning engages students in situations that engineers will encounter in their profession, and includes not only design-implement projects, but also case studies, simulations, and role playing.
- Active learning, which engages students in manipulating, applying, and evaluating ideas, can be applied not only in experiential situations, but also in traditional disciplinary courses and larger class settings.

Studies indicate that students are more likely to achieve intended outcomes and are more satisfied with their education when they are engaged in these kind of learning methods.

This chapter begins by reviewing the results of studies, conducted in CDIO programs, which summarize students' perspectives on their learning. It then outlines an approach to creating active and experiential learning that builds on the curriculum design process. Examples of active and experiential learning illustrate how skills can be integrated into lecture-based courses and design-implement projects. Finally, some key challenges to effective teaching and learning are addressed, including the need for support to enhance faculty competence in teaching and learning. This challenge is addressed again in Chapter Eight.

CHAPTER OBJECTIVES

This chapter is designed so that you will be able to

- appreciate the importance of student perspectives on teaching and learning
- explain the benefits and challenges of integrating skills with engineering disciplinary knowledge outcomes
- describe methods and resources that promote integrated learning
- recognize the importance of aligning curriculum, teaching and learning, and assessment
- give examples of active and experiential learning methods that foster deep understanding of disciplinary knowledge and acquisition of personal and interpersonal skills, and product, process, and system building skills
- appreciate differences in student learning preferences and learning styles

STUDENT PERSPECTIVES ON TEACHING AND LEARNING

Adapting and implementing the CDIO approach in a university engineering program calls for evolving approaches to teaching, learning, and assessment. In the planning stages, it is helpful to get input from students on their experiences with existing learning methods. For example, student representatives at

Chalmers University of Technology, the Royal Institute of Technology, and Linköping University conducted interviews with 56 of their fellow students [1]. Their aim was to contribute student perspectives of teaching and learning during the planning phases that preceded implementation of a CDIO approach. Such interviews have the potential to identify common problematic phenomena regarding student learning. All interview transcripts were interpreted by the student representatives together with experts on teaching and learning, drawing on the research literature on student learning. The study provided the CDIO Initiative with valuable insight into students' perceptions of teaching and learning. Moreover, it validated for student stakeholder groups the appropriateness and necessity of reforming their programs. Table 6.1

TABLE 6.1. STUDENTS' RECOMMENDATIONS FOR MORE EFFECTIVE TEACHING AND LEARNING

1	**Set clear intended learning outcomes relevant to engineering practice.** Clear intended learning outcomes increase motivation and guide studies. Seeing how the course contributes to professional competence is motivating.
2	**Develop teaching activities and assessment tasks that help students reach the intended learning outcomes.** Motivation is increased when students know why they are asked to engage in learning and assessment activities.
3	**Focus on deep working knowledge of basic concepts and provide connections to engineering practice.** This focus promotes a deep approach to learning, increases motivation, and fosters long-term retention.
4	**Prioritize among course content.** Remember that coverage is the enemy of understanding. Reorganizing and reducing content coverage promotes a deep approach to learning and makes for clearer connections among related concepts.
5	**Set an assessment task early in the course.** This helps students get started and gives them an opportunity for an early success. Timely and effective feedback promotes learning and knowing what is expected motivates students.
6	**Set assessment tasks regularly during the course.** Regular feedback is necessary for student learning. Regular monitoring of progress helps students allocate time and keep up with the pace of the course.
7	**Establish explicit criteria for assessment.** Explicit criteria help students focus on the critical aspects of a learning activity or assessment task.
8	**Design learning activities with built-in interaction—both peer interaction and faculty-student interaction.** Interaction is a form of active learning, which in turn is a factor that encourages a deep approach to learning.
9	**Make a realistic plan for time requirements in the course and get regular feedback from students on actual time spent. Coordinate deadlines and workload with parallel courses.** Management of time requirements helps reduce stress levels students experience related to time management issues.
10	**Show enthusiasm for the course and its associated learning tasks.** Faculty enthusiasm enhances the value of the course, and encourages students to appreciate its relevance and worth.

summarizes the results of the study in the form of recommendations for more effective teaching and learning.

Not surprisingly, many of the students' recommendations are related to learning assessment and expectations placed on them. Learning and assessment are, in fact, intertwined. The alignment of assessment with active and experiential learning is addressed more fully in Chapter Seven.

One particularly interesting finding of the same survey is that many students expressed concerns about the usefulness and practical applications of theoretical knowledge. Students frequently felt that engineering theory needed to be memorized for exams, but they could not see connections of theory with real engineering and problem solving. This view is, of course, in stark contrast to the way that faculty see theory as the basis for solving problems and understanding the world around us. Quotations from the student interviews illustrate their views:

- *What the teachers want to know is if you've studied the theory. You often study for the exam and then forget. Instead of focusing on why and how you do something, it is much rote learning.*
- *We should move the focus more toward application—to get a grip on what it's all about. I don't feel that I can apply the knowledge I have.*
- *I want to see the practical use before theory, because that motivates the theory.*

Students point out that in response to the perceived demands of the course (rote learning of theory for the exam), their studying leads to superficial understanding, poor long-term retention, and low motivation. This indicates that many students are adopting a *surface approach to learning* [2]–[4], meaning that students' intention is merely to be able to reproduce the material in order to pass the course. In Chapter Two, factors associated with a surface approach are compared to factors that encourage students to adopt a deep approach to learning. (See Table 2.2) As the students themselves have observed, knowledge resulting from a surface approach is poorly structured and quickly forgotten. The opposite is a *deep approach* to learning, one in which the student's intention is to understand the material. Here, the resulting learning is well structured and tends to be retained in the long term.

The concepts of surface and deep approach to learning are important to bear in mind when designing student learning activities. For a majority of students, the road to understanding and the motivation to learn theory come through applications and connections to real-world problems. When students learn through practical applications, they are more motivated, and they see their education as useful and relevant. This increase in motivation leads to an increase in confidence in their knowledge and skills. As a result, they feel competent and better prepared for their future roles as engineers.

The CDIO Initiative takes students' views into account when planning effective teaching, learning, and assessment methods. The next section of this chapter describes methods to integrate the teaching of disciplinary content with the personal and interpersonal skills, and product, process, and system building skills outlined in the CDIO Syllabus.

INTEGRATED LEARNING

Integrated learning is a key feature of a CDIO program in that students learn personal and interpersonal skills, and product, process, and system building skills together with disciplinary knowledge in the context of professional engineering practice. With integrated learning experiences, faculty can be more effective in helping students apply disciplinary knowledge to engineering practice, and faculty can better prepare students to meet the demands of the engineering profession. Integrated learning allows for dual use of student learning time.

While Standard 3, Integrated Curriculum, emphasizes a systematic plan to integrate skills learning outcomes into a program, Standard 7, Integrated Learning, focuses on the implementation of that plan in each of the program's courses and co-curricular activities. Standards 3 and 7 can be seen as two sides of the same coin.

CDIO STANDARD 7 – **INTEGRATED LEARNING EXPERIENCES**

Integrated learning experiences that lead to the acquisition of disciplinary knowledge, as well as personal and interpersonal skills, and product, process, and system building skills.

Benefits of integrated learning

Integrated learning means that students learn and practice personal and interpersonal, and product, process, and system building skills while learning technical and disciplinary knowledge. Dual-purpose activities serve as vehicles for learning skills while at the same time deepening students' understanding of disciplinary knowledge. Technical knowledge and the learning outcomes related to the CDIO Syllabus are interdependent and developed together. For example, in a CDIO program, communication skills are deeply embedded in technical knowledge. Students acquire the ability to communicate technically, both with experts and non-experts. They gain confidence in expressing themselves within their field of work. They are expected to be able to describe and present ideas, argue for or against conceptual ideas and solutions, and develop ideas through collaborative sketching and engineering reasoning. It is obvious that these communication skills are inseparable from students' expression and application of technical knowledge. Learning activities and assessment should therefore be modified to address learning outcomes related to disciplinary knowledge simultaneously with skills. Learning and assessing communication is most effective in authentic contexts, that is, in situations that simulate engineering practice.

Practicing skills within engineering contexts also enables students to acquire a deeper working knowledge of engineering fundamentals. To make dual use of learning time, learning activities and assessment methods must adopt new approaches. It is important to note that integrating skills learning outcomes into a course is not about adding a lot of new theoretical content

Box 6.1. Integrated Learning of Communication Skills

In the Mechanical Engineering program at Chalmers University of Technology, oral and written communication skills are integrated into three courses in the first three years of the curriculum: the introductory course in the first year, a design-implement project course in the second year, and as an integral part of the thesis work at the end of the third year. During the first three years, the development of communication skills is focused mainly on academic writing, even though there are strong elements of reflective writing, that is, writing in order to learn. On the Master's level (4th and 5th years of the program), communication skills are further developed with emphasis on improving learning of the technical content.

The intended learning outcomes are that students should be able to write both technical (design) reports and scientific reports, to give oral presentations using presentation tools, and to be able to make poster presentations. Project reports also include presentations of course projects and assignments. From the table, we can see that assessment is carried out through feedback on the different activities, where language and communication teachers are working together with engineering faculty to assess content, form, and language. On the Master's level (4th and 5th years), communication skills are integrated mainly into project-based courses. However, even in some lecture-based courses, we highlight the importance of effective communication. For example, in a course on Internal Combustion Engines, students give oral reports of their assignments. Feedback is given on the presentation, both on the delivery, and the quality and relevance of slides.

	Introductory Course (1st year)
Integrated task	Write a technical report and give an oral presentation of the project assignment.
Lectures	How to write the technical reports; oral presentations; multimedia and electronic communication
Discussion	Communication and critical thinking; writing as a methods for reflection; form and content of written presentations
Exercise	Graphic communication skills; sketching
Feedback	Feedback on written reports and oral presentations
	Project Course (3rd year)
Lectures and discussion	Communication strategy; multimedia; written communication
Feedback	Feedback on oral presentations
	Thesis Work (3rd year)
Lectures and discussion	Critical evaluation of scientific information; developing information literacy; how to write scientific papers; how to make a poster
Feedback	Troubleshooting and feedback sessions; feedback on reports and posters; feedback on oral presentations

– S. Andersson, Chalmers University of Technolog

into an already densely packed course. The Syllabus is not to be interpreted as the table of contents for a whole new body of theoretical knowledge from the disciplines of sociology, psychology, philosophy, or economics. Rather, it lists important aspects of applying technical knowledge as a professional engineer. They are skills, or competencies, and as such they should be developed through cycles of application, feedback and reflection.

Integrated learning across multiple experiences

Some skills, such as teamwork and communication skills, need to be taught and assessed in several courses throughout the program. Learning activities are sequenced to build upon students' previous experiences rather than starting over in each course or learning experience. In Chapter Four, the curriculum design process addresses the placement of personal and interpersonal skills, and product, process, and system building skills in sequenced courses. (See Figure 4.7) This same sequence forms a framework for planning integrated teaching and learning activities. For example, methods of teaching written communication skills can be coordinated across several courses, even when different faculty members teach written communication. Box 6.1 gives an example of the integration of communication skills into the teaching and learning of the disciplinary content in a mechanical engineering program at Chalmers University of Technology.

METHODS AND RESOURCES THAT PROMOTE INTEGRATED LEARNING

Planning for integrated learning begins by deciding on the purpose of the course. This is done by specifying the intended learning outcomes. The learning outcomes of a course include not only disciplinary content, but also learning outcomes related to the CDIO Syllabus. These may be partly specified in the curriculum design process, as described in Chapter Four. The described curriculum design process will go as far as ensuring that the outcomes of the Syllabus are covered in the curriculum in approximately the right frequency and sequence. However, refinement and detailing of the outcomes is the responsibility of each course. Explicitly specifying skills in the course learning outcomes helps to ensure that they will be taught and assessed. Conflicts could otherwise arise when faculty disagree on the purpose of the course. For example, the development of teamwork skills may be seen as a secondary side effect that may or may not happen in a course where students work in teams. Alternately, teamwork skills can be seen as an important outcome of the course, an outcome which must be carefully addressed in course development. The process of explicitly defining and agreeing on the intended learning outcomes is a way to resolve the issue and avoid unnecessary conflict. The specification of learning outcomes began in Chapter Three.

Specification of intended learning outcomes

Intended learning outcomes describe what the student will be able to do as a result of participating in the course. Statements of student performance use active cognitive verbs and verb phrases, such as *describe, give examples of, choose, explain in your own words, estimate, calculate, solve, apply, design, interpret, plan, evaluate, modify, decide, sketch, critique*. These outcomes should be met by performances that are observable, that is, it must be possible for students to demonstrate, and faculty to decide, whether the outcomes have been met. Assessable learning outcomes are written by connecting a topic from the CDIO Syllabus to a cognitive verb that indicates the desired proficiency level. Examples of such assessable learning outcomes include *"Discuss and determine the statistical validity of data"* and *"Elicit and interpret customer needs"*. Collecting evidence of demonstrated performance of the intended learning outcomes is the focus of student learning assessment and is discussed in Chapter Seven.

Intended learning outcomes also point to the level of understanding or skill that students must reach. As discussed throughout this book, engineering education in CDIO programs should result in engineers who can conceive, design, implement, and operate complex value-added engineering systems in a modern team-based environment. This means that students must acquire conceptual understanding of engineering theory and principles, and be able to analyze, apply, and evaluate these concepts.

Classification of intended learning outcomes

Classifications of intended learning outcomes, or taxonomies, are useful tools for specifying learning outcomes aimed at different levels of understanding. Teaching and learning activities, as well as assessment methods, should be aligned with the intended level of understanding. As described in Chapter Three, Bloom's *Taxonomy of Educational Objectives* [5] lists six levels of understanding: Knowledge, Comprehension, Application, Analysis, Synthesis, and Evaluation. Bloom and his colleagues suggested a hierarchical framework from knowledge to evaluation, with each level subsuming the previous ones. (See Table 3.8 for examples of learning outcomes aligned with Bloom's Taxonomy.)

For technical disciplinary courses that focus on problem solving with calculations, the Feisel-Schmitz *Technical Taxonomy* [6], may also be useful. (See Table 6.2.) The five levels of understanding of the Feisel-Schmitz *Technical Taxonomy* include Define, Compute, Explain, Solve, and Judge. Each classification level lists specific learning outcomes. This taxonomy makes a useful distinction between two levels of problem solving, *Compute* and *Solve*. *Compute* means following a known procedure to solve a standard problem. *Solve* refers to a higher level of problem solving in which an element of modeling, or synthesis of knowledge, is required, and thus a higher level of conceptual understanding is necessary. The hierarchical nature of the taxonomy means that students can reach the level of *Compute* without being able to *Explain*, while *Solve* includes all the underlying levels *Explain, Compute,* and *Define*.

TABLE 6.2. FEISEL-SCHMITZ TAXONOMY [6]

Feisel-Schmitz Taxonomy	Sample learning outcomes
Define	State the definition of the concept or describe in a qualitative or quantitative manner
Compute	Follow rules and procedures; substitute quantities correctly in equations
Explain	State the concept in one's own words; explain the procedure used; discuss the outcome
Solve	Characterize, analyze, or synthesize to model a system; modify a model of a system; provide assumptions
Judge	Critically evaluate multiple solutions; select an optimum solution; evaluate supporting evidence

TABLE 6.3. EXAMPLES OF INTENDED LEARNING OUTCOMES RELATED TO SPECIFIC TOPICS IN THE CDIO SYLLABUS

As a result of this learning experience, you should be able to . . .	Relates to CDIO Syllabus:
Explain, at a level understandable by a non-technical person, how jet propulsion works	1.3 Advanced Engineering Fundamental Knowledge
Compare experimental data to available models	2.2.3 Experimental Inquiry
Formulate solutions to problems using creativity and effective decision making skills	2.4.3 Creative Thinking
Analyze the strengths and weaknesses of the team	3.1.2 Team Operation
Use appropriate nonverbal communication, for example, gestures, eye contact, poise, when giving oral presentations	3.2.6 Oral Presentation and Interpersonal Communications
Appraise the operation system for your team's product and recommend improvements	4.6.4 System Improvement and Evolution

Stating intended learning outcomes at specific levels is the basis for choosing appropriate teaching, learning, and assessment methods.

Examples of intended learning outcomes

Course-level learning outcomes should be based on a locally customized CDIO Syllabus, and the integrated curriculum design. Intended learning outcomes should be stated in terms of observable performances and should indicate the level of understanding students are meant to demonstrate. All learning outcomes for a course should be realistic with regard to student time and resources, and should be explicit in that they are clear to faculty, students, and other stakeholders. Table 6.3 gives examples of CDIO learning outcomes that are used in CDIO programs and their relationship to a hypothetical set of learning outcomes, based on the CDIO Syllabus, that might be assigned to a course.

Constructive alignment of intended learning outcomes

Integrating personal and interpersonal skills, and product, process, and system building skills into a course means that they are explicitly addressed in the learn-

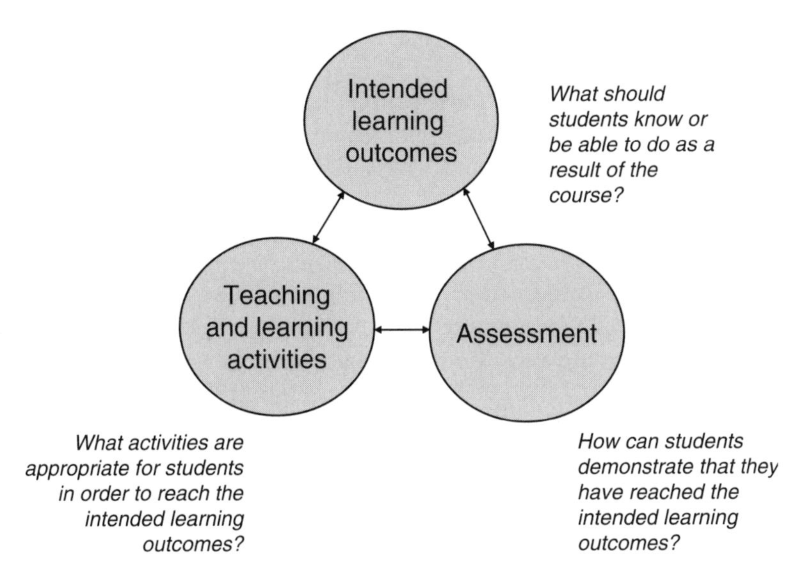

FIGURE 6.1. **CONSTRUCTIVE ALIGNMENT OF OUTCOMES, TEACHING AND LEARNING, AND ASSESSMENT**

ing outcomes of the course, in the teaching-learning activities, and in the assessment of student learning. This purposeful relationship between the intended learning outcomes, teaching-learning activities, and assessment is known as *constructive alignment* [4], as illustrated in Figure 6.1. In this chapter, we focus on the alignment of teaching-learning activities with the intended learning outcomes of a CDIO course. Chapter 7 examines the alignment of assessment methods with intended learning outcomes and with teaching and learning.

Many learning outcomes represent knowledge, skills, and attitudes learned primarily through application and practice. Theoretical knowledge alone is not sufficient. Skills learning outcomes should be deliberately planned and taught. For example, asking students to work in teams does not automatically mean that they will learn effective teamwork skills. Issues such as how to form a team, how to plan and apportion work within a team, and how to resolve conflict within a team must be explicitly addressed. Productive learning occurs when activities give students specific opportunities for practice, reflection on their experiences, and applications of theoretical concepts. In order to meet the challenge of designing integrated learning experiences, we need to explore the benefits of active and experiential learning. Integrating all learning outcomes provides opportunities for faculty to design authentic learning tasks, that is, to engage students in the same kinds of tasks that engineers perform.

Faculty support for integrated learning

If personal and interpersonal skills, and product, process, and system building skills are an essential component of engineering practice and education, they need to be integrated into each course in the curriculum. Students must be able to recognize faculty as professional engineers, instructing them in both disciplinary content and skills. To redevelop their courses to include active and experiential learning, faculty need opportunities to improve their teaching and assessment skills. Many departments have a deeply rooted culture of conventionally taught lecture courses, and may quickly discover that it takes a great deal of effort to increase the use of, and competence in, new teaching and assessment methods. Planning for active and experiential learning requires time, resources, and support from experts in learning and assessment.

While faculty in CDIO programs are expert in their engineering disciplines, in many cases, they have relatively little background in the knowledge and skills beyond the first section of the CDIO Syllabus, which describes disciplinary knowledge. In addition, resource constraints frequently limit the hiring of experts who can design experiences that address all of the desired skills. If instruction in knowledge, skills, and attitudes is to be integrated into existing disciplinary instruction, faculty will need support to augment their teaching and assessment activities. When faculty teach skills as part of their disciplinary teaching-learning activities, students see them as important to engineering faculty. In addition, students understand that these skills are a critical part of their success as engineers.

In order to facilitate the design of integrated teaching and learning activities, resources for effective practice, including examples, have been gathered and organized in a way that reflects the organization of the CDIO Syllabus. These resources, called *Instructor Resource Materials* (IRM), address the constraints on implementing integrated curriculum, teaching, and assessment with examples and resources for specific skills. Each IRM provides instructional resources, teaching suggestions, and assessment tools for a specific skill area that engineering faculty worldwide can use to integrate skills learning outcomes into their existing engineering course materials. It is important to emphasize that these materials are for instructors, not students, to use in developing integrated learning experiences. They provide ideas and materials that faculty can adapt for their courses. In addition to specific teaching and learning materials, there are references to other articles, websites, books, and experts who can provide additional information. IRMs include examples of best practices where the material has been used in other classroom situations. As part of a continuous improvement process, IRM designers solicit materials and feedback from experts who often wish to contribute supplemental material.

ACTIVE AND EXPERIENTIAL LEARNING

Active learning methods engage students directly in thinking and problem-solving activities. There is less emphasis on passive transmission of information and more emphasis on engaging students in manipulating, applying, analyzing, and evaluating ideas. By engaging students in thinking about concepts, particularly new ideas, and requiring some kind of overt response, students not only learn more, they recognize for themselves what and how they learn. This process helps to increase students' motivation to achieve intended learning outcomes and form habits of lifelong learning. Active learning is considered *experiential* when students take on roles that simulate professional engineering practice, such as, design-implement projects, simulations, and case studies. CDIO Standard 8 addresses the need for active and experiential learning.

CDIO STANDARD 8 – **ACTIVE LEARNING**

Teaching and learning based on active and experiential learning methods.

Active learning is known to support a deep approach to learning [3]-[4]. As explained earlier in this chapter, a deep approach to learning means that students intend to understand the concepts, as opposed to simply reproducing the information on an exam. Active and experiential learning methods influence the approach that students are likely to adopt. When students are given an active role in their learning process, they learn better because they are more likely to take a deep approach to learning. Students who are actively involved in their own learning make better connections, both with past learning and between new concepts [4].

Active learning methods

With active learning methods, faculty can help students make connections among key concepts and facilitate the application of this knowledge to new settings. These methods can be incorporated into all types of courses. Methods that are suitable for active learning in lectures include muddy cards, concept questions, electronic response systems, ticking, discussions with partners or small groups, and variations of these methods [7]-[10]. A few of the more widely used methods in CDIO programs are described here.

Muddy cards. Muddy cards, also known as *Muddiest-Point-of-the-Lecture* cards, gather in-class feedback to determine gaps in student comprehension [9]. Near the end of a lecture or other learning experience, students are asked to reflect on what they have learned. They write down the concepts or ideas—the *point*—they found most unclear—the *muddiest*. The instructor collects the cards for later review. Muddy points can be addressed in a number of ways: posting questions and answers on the course website, answering questions at the start of next class meeting, distributing printed copies of answers to the most common muddy points, or sending email responses to the class. Faculty who use muddy cards have experienced many benefits from the cards. Muddy cards provide time for student reflection, with a consequent increase in learning retention. Writing their questions and comments helps students to organize their thoughts and study more effectively. Moreover, the cards provide information to the instructor in time to correct misconceptions by the next class meeting and to assist in improving the course for the next offering. Despite the fact that muddy cards can require a large commitment of time—particularly the first time they are used—many faculty have found them very useful both as an active learning method and as an assessment method.

Concept questions. Concept questions are multiple-choice items used to gather in-class feedback to check for student understanding and to correct student misconceptions [10]. Because of their format, concept questions require only a few minutes for thinking and responding. Questions focus on a single concept, are not solvable by relying solely on equations, reveal students' common difficulties with the concepts, and have more than one plausible answer. This active learning method requires that the instructor develop questions while preparing the lecture, and pose those questions at appropriate points while delivering the lecture. A question is displayed with its answer choices, and students are given a few minutes to respond. Students indicate their responses with raised hands, color-coded index cards, or electronic systems. Figure 6.2 illustrates two examples of concept questions used in a thermodynamics course.

An airplane powered by two jet engines accelerates during takeoff. Assume the exhaust velocity relative to the vehicle is constant and the mass flow into the engine is constant. Neglecting any forces due to the acceleration of the vehicle, how does the thrust vary as the aircraft accelerates?

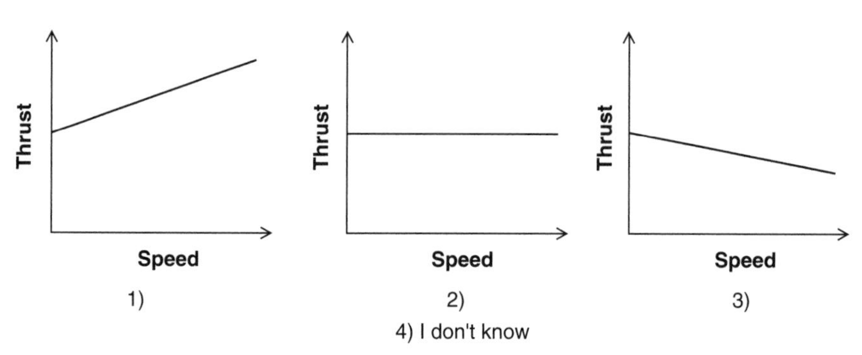

AIRCRAFT ENGINE PERFORMANCE

To maximize endurance, an airplane must flow in a manner that

1. **minimizes drag**
2. **maximizes drag**
3. **maximizes lift/drag ratio**
4. **maximizes power available**
5. **minimizes power required**
6. **I don't know**

-- I. Waitz, Massachusetts Institute Of Technology

Figure 6.2. Sample Concept Questions in Thermodynamics

Most CDIO programs use concept questions to enhance student learning in lectures. When students seem not to understand a concept, they are asked to discuss the alternative responses with a partner. This *peer instruction*, advocated by Eric Mazur [10], has been found to be effective across a broad range of teaching applications. Faculty find that using concept questions gives them an indication of student understanding and helps them to adjust the lecture in real time to correct student misunderstandings. Students appreciate concept questions because they get feedback that helps them to plan study time. As with the use of muddy cards, concept questions require additional time for preparation and execution, but faculty and students alike find them to be effective means toward helping students achieve deeper conceptual understanding.

Electronic response systems. Some of our programs use an electronic response system, such as the *Personal Response System* (PRS), in conjunction with concept questions [11]. An electronic response system is a way to collect, summarize, and display in-class feedback. With infrared wireless transmitters that look like television remote controls, students can indicate their responses with the click of a button. An electronic response system offers an anonymous method for collecting feedback simultaneously from all students. Results are instantly graphed and can be displayed for the instructor only, or for the whole class. Faculty have found that electronic response systems increase student motivation and class participation. With the software provided, responses can be archived and analyzed at a later date to help with future course planning and assessment. Though not prohibitive, there is a cost to equipping classrooms with these systems.

Ticking. Ticking is an active learning method appropriate in many basic engineering courses where the focus is on problem solving. It is used at the Royal Institute of Technology (KTH) in Stockholm to enhance recitations sessions and large-group tutoring sessions [12]. For each recitation session, students are asked to work through a set of problems. At the recitation, students *tick* on a list which problems they are willing and prepared to present. The instructor randomly selects a student from the list to present the first problem on the board, followed by a second student to lead the next problem, and so on. Ticking at least 75% of the problems in the course is rewarded with bonus points for the exam or a similar reward. The reward is given for the ticks themselves, not for the quality of the presentations. There is a minimum requirement, however, that students must demonstrate that they have made an honest effort in preparing the problem. They must be able to lead a classroom discussion to a satisfactory solution. If students cannot achieve this goal, their ticks are removed for that recitation session. At KTH, students react positively to this active learning method, and they often comment in course evaluations that the method helps them to learn.

Ticking promotes effective learning for several reasons:

- Preparing the weekly problems motivates students to spend time on the task. For every problem presented by one student, 30 times as many have prepared the problem. Also, attendance at recitations is likely to increase.
- Ticking generates appropriate learning activity. Because they must not only solve the problems, but also be prepared to present them in class, students are encouraged to reflect on how to explain their methods and decisions.
- Because students have prepared exactly the same problems that others present, they are able to follow the problem-solving approaches, even when the presentation is less than perfect. This active learning method provides immediate feedback for all students, and often leads to good discussions on alternative solutions.

In addition, ticking takes advantage of a social dimension that creates strong student motivation. It is important that students perceive the emotional climate as safe and friendly.

Experiential learning methods

As defined earlier, experiential learning engages students by setting teaching and learning in contexts that simulate engineering roles and practice. Experiential learning methods include project-based learning, simulations, case studies, and the design-implement experiences which are the focus of Chapter Five. These methods are based on pedagogical theories of how students, especially engineering students, learn and develop cognitive skills.

The CDIO approach to engineering education is based on experiential learning theory. The learning cycle, proposed by Kolb [13], provides helpful insights for planning teaching and learning activities. One such planning application is illustrated in Figure 6.3 [14].

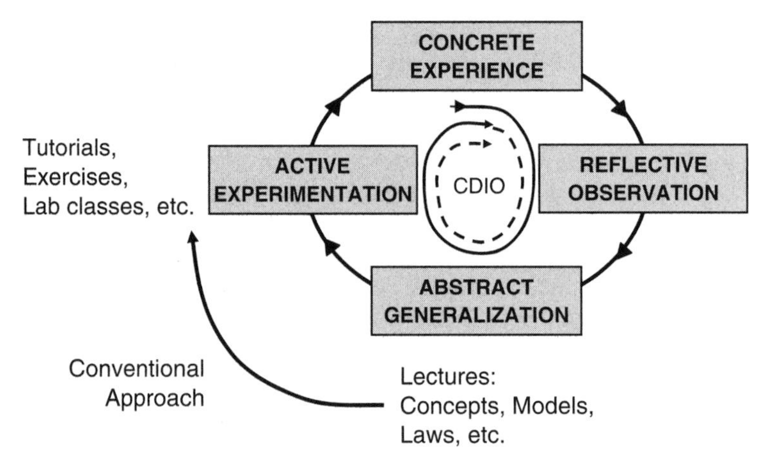

FIGURE 6.3. EXPERIENTIAL LEARNING MODEL (ADAPTED FROM KOLB, 1984) USED WITH PERMISSION OF PRENTICE-HALL.

In a CDIO program, the experiential learning cycle is entered at different points. Lecture-based courses that incorporate active learning begin with reflective observation to stimulate learning because students have a common base of experience. Lectures may also begin with abstract generalization and conclude with active experimentation, for example, problem sets or exercises. As a result, students of various learning styles are accommodated. The introductory engineering course provides the first concrete experience in engineering, creating the cognitive framework for subsequent learning of theory. In design-implement experiences, concrete experiences are the entry point to the experiential learning cycle. Students engage in tasks similar to real engineering practice, reflect on what they have learned from these experiences, generalize their learning to develop abstract ideas and principles, and test these new ideas with active experimentation and application to other problems. Embedded experiential learning throughout the curriculum provides opportunities for reinforcement. The capstone experience provides opportunities to apply theory and to build students' confidence in engineering.

Project-based learning. Project-based learning is built on an authentic, or real world, situation or problem for which a solution is found. That solution may include design-implement experiences. For the most part, we do not follow a problem-based project-organized curriculum, as does, for example, Aalborg University in Denmark [15]. However, it is common to use project-based learning activities in an existing curriculum framework. Faculty identify problems that encompass the concepts and principles relevant to the content domain, and they design authentic tasks in which the thinking required is consistent with the thinking in an engineering environment. The task and environment reflect the complexity of engineering environments and encourage students to test their ideas against alternative views and contexts. Projects provide opportunities for reflection on both the content learned and the learning process.

Faculty have found that project-based learning increases student motivation and improves students' ability to apply engineering knowledge and skills to real-world problems. However, the resources required to design and monitor learning experiences—people, time, equipment and space—may place some limitations on the use of project-based learning. In addition, this approach requires a change in faculty role, from lecturer to coach or mentor. Our experiences with project-based learning, including recommendations for addressing its major limitations, are outlined in a report that is available on the CDIO website [16].

Simulations. Similar to project-based learning, simulations are activities in which students take on engineer-like roles in the application of engineering laws or principles. Simulations often have specific rules, guiding principles, and structured roles and relationships [17]. The instructor's role in a simulation is to explain the rules, the situation, and the roles students are to take on; to monitor the simulation as it is played out, to help students reflect on the experience; and to lead a debriefing session. Most simulations are based on computer hardware and software. For example, the aeronautics and astronautics program at the Massachusetts Institute of Technology (MIT) uses a flight simulator to give students practice in piloting in preparation for the design-implement-fly contest of a radio-controlled plane. Faculty who use simulations have found that these learning activities provide students with opportunities to experience engineering tasks in a safer environment than the real situation would require. In addition, students have access to what would otherwise be scarce or inaccessible equipment and facilities.

Case studies. Although case studies have been used primarily in law, business, and medical education, they are equally appropriate for engineering education. A good case tells the story of a real engineering experience, usually from the point of view of the participants. In addition to the narrative, a typical case provides detailed background, such as, original calculations and drawings, budget and schedule limitations, the availability of material resources and technical facilities, and the people and organizations involved in accomplishing the task [18]. Through a case discussion, students vicariously experience the activity in the case and are involved in the resolution of problems and issues. The goal is to help students develop independent thinking and decision-making skills through practice. Faculty who use case studies have found that this teaching and learning technique helps students to develop analytic and problem-solving skills, and enables them to explore solutions to complex issues. Moreover, the glimpse at what engineers do provides background information on the history and traditions of the engineering profession. One difficulty that faculty face with the case study approach is finding cases relevant to specific disciplinary content.

Using multiple active and experiential methods

Many faculty combine two or more active and experiential learning methods in a single course. For example, an advanced course in aerodynamics at MIT combines four methods: concept questions, an electronic response system, readings and problems assigned prior to lecture, and team-based project-based learning. In addition, the course includes oral examinations as a method of student learning assessment. Box 6.2 describes this example.

Making engineering education attractive to all students

CDIO programs integrate the learning of personal and interpersonal skills, and product, process, and system building skills with disciplinary knowledge, through active and experiential learning methods. Because of this approach to learning, engineering is more attractive to students who have not traditionally chosen engineering as a field of study or as a career. Although improving, engineering education has a long tradition of being a predominantly male environment, with underrepresentation of women and ethnic minorities. In selecting active and experiential learning methods, we need to be sensitive to the ways in which we can help all students to thrive and succeed.

Box 6.2. Active And Experiential Learning in an Aerodynamics Course

Active and experiential learning methods have transformed an advanced course in aerodynamics at the Massachusetts Institute of Technology. *Aerodynamics* is a 3rd-yr undergraduate course with a typical enrollment of 40 students. Prior to 1999, the course was a traditional engineering course with lectures, recitations, weekly homework assignments, a small end-of-semester design project, and written exams. The current course includes several active and experiential learning activities:

- *Concept-based lectures with real-time feedback*
  In this approach, two or three multiple-choice concept questions are part of lecture. These questions are designed to include the important concepts of the subject and their common misconceptions. After a few minutes of independent reflection, students use an electronic response system to select an answer. Responses are charted and projected in real-time. Depending on the responses, students discuss their answers with each other or the instructor clarifies misconceptions.
- *Weekly (graded) homework given prior to class lecture and discussion*
  In *Aerodynamics*, students complete homework assignments and related readings prior to in-class discussion. With this preparation, the classroom becomes an interactive environment where students bring a common language to discuss the conceptual difficulties they have encountered.
- *Semester-long, team-based project involving analysis and design of an aircraft*
  In this project-based approach, theoretical knowledge is immediately applied to the complex design of modern aircraft. In addition, the use of a semester-long project provides a context for learning the technical fundamentals. In recent years, two design projects have been developed, one based on a military fighter aircraft, and another on a blended-wing body commercial transport aircraft.
- *Oral examinations*
  Oral examinations take an active approach to assessment of student learning. They provide insight into how students understand and relate concepts. Furthermore, practicing engineers are faced daily with the real-time need to apply rational arguments based on fundamental concepts. By using oral exams, it is possible to assess a student's ability to construct sound conceptual arguments.

End-of-semester student evaluations of the changes in pedagogy and assessment reveal these findings:

- The new pedagogy is consistently rated as highly effective.
- Challenging pre-class homework increases the effectiveness of lecture.
- An increase in student learning occurs over the length of the semester, as students transition to the new approaches.
- Effective implementation of the team project is difficult.
- Oral examinations are effective in helping the instructor to determine if students have achieved the course learning outcomes.
- Many students find oral exams to be a more accurate representation of their understanding than traditional written exams. In fact, several students have said that the oral exams were the best parts of the course.

– D. Darmofal, Massachusetts Institute of Technology

Research studies on the learning preferences of students from underrepresented populations identify specific factors that are important to them in their learning environments. In Sweden, for example, a study conducted with female engineering students revealed that, when compared with their male counterparts, female students prefer:

- personal contact with faculty
- regular feedback on their work

BOX 6.3. GENDER DIFFERENCES IN LEARNING PREFERENCES

In two programs at Linköping University, *Computer Science* and *Applied Physics and Electrical Engineering*, female students represent fewer than 10% of all students. In a project aimed at making students aware of gender differences in learning styles, female students were asked to observe lectures, lessons, project work, examinations, and course readings from a gender perspective. Eleven students at different academic levels volunteered for one semester. They were given an introduction to gender studies from researchers in the field. They also attended regular meetings with tutors who helped them reflect on their experiences and discuss them with other observers. Four main findings resulted:

- *There is a shortage of female role models*
  There are few female faculty. Apart from a few non-technical courses, some students had not met a female lecturer until their third year. There were also few female authors in the course texts and readings. In one of the engineering programs, only four books out of 90 were written by women.
- *Engineering examples represent a male perspective*
  One project task involved the development of a computer game that contained violence. The female students felt they could not relate to this sort of adventure game and believed that the university should not allow this kind of computer game. In some courses, male faculty explained the application of a theoretical problem through his personal experience in military service, one that is not familiar to most female students in Sweden.
- *Roles in student teams are affected by gender*
  Female students took the role of secretary more often than team leader. Female students also considered it important to have more than one female in a project group, at least in the first two years of the program.
- *Female students seem to be more affected by poor teaching than are their male counterparts*
  The reasons for this outcome are not entirely clear, and warrant better definitions of *good* and *poor* teaching, as well as further investigation.

The results were consistent with research in gender studies. In addition, the group discussions played an important part. Students found that being able to reflect upon their experiences and the structures in their learning environments led to deeper insights and maturity. Since the study, students have found themselves using gender perspectives even outside the university. This project has led to students' personal growth and critical thinking—qualities that are key goals of universities. Increasing awareness of gender differences in learning preferences can lead to positive changes in academic programs. Among students' recommendations are more female role models, assigning mentors from senior classes to new female students, and educating faculty and students about gender differences in teaching and learning.

– M. ENGSTRÖM, LINKÖPING UNIVERSITY

- assignments that are relevant to their lives
- teaching and learning set in comprehensive contexts
- effective time management in project work
- assessment by performance rather than by traditional examination
- interaction with appropriate role models [19]

A related study of learning preferences, conducted at Linköping University, is summarized in Box 6.3. Changes brought about by CDIO programs would be positive steps toward an engineering education where all students can flourish.

BENEFITS AND CHALLENGES

We have been studying the results of these innovations in teaching and learning methods. As a result of integrating skills with disciplinary content, and using active and experiential learning, we find that

- Introducing the CDIO approach has deepened—not diminished—students' understanding of engineering disciplines.
- Annual surveys of graduating students indicate that they have developed intended CDIO Syllabus knowledge and skills.
- Course evaluation results indicate that faculty are using a wider variety of teaching and assessment methods than they were previously.
- Student self-report data indicate high student satisfaction with their learning experiences.
- Longitudinal studies of students in CDIO programs show increases in program enrollment, decreasing failure rates, particularly among female students, and increased student satisfaction with learning.

These benefits will be examined in more detail in Chapter Nine.

Integration of personal and interpersonal skills, and product, process, and system building skills with disciplinary content, and active and experiential learning methods are not, however, without challenges.

- Despite evidence to the contrary, faculty sometimes perceive a conflict between disciplinary content and the learning of personal and interpersonal skills, and product, process, and system building skills.
- Faculty are sometimes reluctant to reduce the amount of material covered in their courses because subsequent courses depend on a full coverage of topics.
- Faculty and students often resist changes to the ways they are accustomed to teaching and learning.
- Faculty may lack the expertise to implement active and experiential learning methods.

The issue of enhancing faculty teaching, learning, and assessment methods is addressed in Chapter Eight.

SUMMARY

Integrated learning makes it possible for students to learn personal and interpersonal skills, and process, product, and system building skills, while simultaneously deepening their conceptual understanding of disciplinary knowledge. Students practice and learn engineering in authentic contexts and are more satisfied with their learning experiences. In 1st- and 2nd-year courses, they are also introduced to, and motivated to learn, disciplinary abstractions. In subsequent years, this disciplinary knowledge is reinforced by application, and the students are empowered by their accomplishments.

Active learning methods, for example, muddy cards, concept questions, and electronic response systems, engage students in their learning by requiring deliberate mental effort and evoking overt responses. Faculty can adapt a variety of active learning methods to lecture-based and project-based courses and seminars. With experiential learning, such as project-based learning, simulations, and case studies, students have opportunities to take on a variety of engineering roles in increasingly complex learning situations.

CDIO programs integrate the learning of personal and interpersonal skills, and product, process, and system building skills with disciplinary knowledge through active and experiential learning methods. Because of this approach to learning, engineering is more attractive to students, and it is especially attractive to those students who traditionally have not chosen engineering as a field of study or career.

Together with Chapters Four and Five, this chapter provides an answer to the second central question for engineering education: *How can we do better at ensuring that students learn these skills?* We focused on learning outcomes that integrate skills with disciplinary knowledge, and the alignment of teaching and learning methods with these outcomes. Chapter Seven continues this approach with the alignment of student learning assessment methods with the learning outcomes and teaching and learning methods.

DISCUSSION QUESTIONS

1. In what ways can you begin to integrate personal and interpersonal skills, and product, process, and system building skills into your courses?
2. What active and experiential learning methods are used effectively in your courses?
3. How can you find out more about your students' perceptions of their learning?
4. How would you begin to address the key challenges to integrated learning and active and experiential learning posed in this chapter?

References

[1] Edström, K., Törnevik, J., Engström, M., and Wiklund, Å., "Student Involvement in Principled Change: Understanding the Student Experience", *Proceedings of the 11th International Symposium Improving Student Learning, OCSLD*, Oxford, England, 2003.

[2] Marton, F., and Säljö, R., "Approaches to Learning", in Marton, F., Hounsell, D., and Entwistle, N. J. (Eds.) *The Experience of Learning*. Edinburgh: Scottish Academic Press, 1984.

[3] Gibbs, G., *Improving the Quality of Student Learning*, TES, Bristol, England, 1992.

[4] Biggs, J., *Teaching for Quality Learning At University*, 2nd ed., The Society for Research into Higher Education and Open University Press, Berkshire, England, 2003.

[5] Bloom, B. S., Englehart, M. D., Furst, E. J., Hill, W. H., and Krathwohl, D. R., *Taxonomy of Educational Objectives: Handbook I—Cognitive Domain*, McKay, New York, 1956.

[6] Feisel, L. D., "Teaching Students to Continue Their Education", *Proceedings of the Frontiers in Education Conference*, 1986.

[7] Angelo, T. A., and Cross, K. P., *Classroom Assessment Techniques: A Handbook for College Teachers*, 2nd ed., Jossey-Bass, San Francisco, California, 1993.

[8] Johnson, D.W., and Johnson, R., *Leading the Cooperative School*. Interaction Book Company, Edina, Minnesota, 1994.

[9] Mosteller, F., 1989, "The 'Muddiest Point in the Lecture' as a Feedback Device", *On Teaching and Learning*, Vol. 3, pp. 10-21. Available at http://isites.harvard.edu/fs/html/icb.topic58474/mosteller.html

[10] Mazur, E., *Peer Instruction: A User's Manual*. Prentice Hall, NJ, 1997.

[11] InterWritePRS. Available at http://www.gtcocalcomp.com/interwriteprs.htm

[12] Lindqvist, K., and Edstrom, K., *Ticking*, The Royal Institute of Technology, Stockholm, Sweden. Personal communication.

[13] Kolb, D. A., *Experiential Learning*, Prentice-Hall, Upper Saddle River, New Jersey, 1984.

[14] Cunningham, G., Queen's University, Belfast, Northern Ireland.

[15] See Aalborg University at http://www.aau.dk

[16] Andersson, S., Edström, K., Eles, P., Knutson Wedel, M., Engström, M., and Soderholm, D., *Recommendations to Address Barriers in CDIO Project-Based Courses*, CDIO Report, 2003. Available at http://www.cdio.org

[17] Bonwell, C., and Eison, J., *Active Learning; Creating Excitement in the Classroom*. ASHE-ERIC Higher Education Report 1. The George Washington University, School of Education and Human Development, Washington, DC, 1991.

[18] Kardos, G., "Engineering Cases in the Classroom", *Proceedings of the National Conference on Engineering Case Studies* 1979. Available at http://www.civeng.carleton.ca/ECL/cclas.html

[19] Salminen-Karlsson, M., *Bringing Women into Computer Engineering*, Linköping University, Department of Education and Psychology Dissertations No 60. Linköping, Sweden, 1999.

CHAPTER SEVEN STUDENT LEARNING ASSESSMENT

WITH P. J. GRAY

INTRODUCTION

The last three chapters have discussed answers to the second of the two questions central to the reform of engineering education: *How can we do better at ensuring that students learn these skills?* Integrated curriculum, design-implement experiences, integrated learning, and active and experiential learning are the main components of a reformed engineering education that better ensures that students reach the intended outcomes required of all engineering graduates. Implicit in the question *"How can we do better . . . "* is an additional question: *How do we know that we are doing better?*

- How do we know that students are achieving the intended learning outcomes?
- How do we know that our engineering programs are effective?

We answer the first part of the question in this chapter on student learning assessment, and then return to address the second part of the question later in Chapter Nine on program evaluation. Student learning assessment measures the extent to which each student achieves specified learning outcomes. Faculty members plan and implement student learning assessment with respect to the outcomes within their courses. In contrast, program evaluation examines the key success factors of CDIO programs in terms of both the overarching student learning outcomes and the adoption of the CDIO Standards.

Learning is assessed before, during, and after instructional activities. Formative assessment collects evidence of student achievement while students are in the process of learning. Results of formative assessment inform students about their progress, help monitor the pace of instruction, and indicate areas of instruction that may need to be changed. Summative assessment gathers evidence at the end of an instructional event, such as a major project, a course, or an entire program. Results of summative assessment indicate the extent to which students have achieved the intended learning outcomes of the project, course, or program. If the instructional event will be repeated with other students, summative assessment, as well as formative assessment, is used to improve curriculum, teaching-learning methods, and the design and use of learning spaces.

Assessment of student learning in personal and interpersonal skills, in process, product, and system building skills, and in disciplinary knowledge has four main phases:

- Specification of learning outcomes
- Alignment of assessment methods with curriculum, learning outcomes, and teaching methods
- Use of a variety of assessment methods to gather evidence of student achievement
- Use of assessment results to improve teaching and learning

The importance of specifying learning outcomes and aligning them with teaching and learning has been highlighted in previous chapters. The focus now is on assessment methods appropriately matched to curriculum and teaching methods. Effective learning assessment is aligned with intended learning outcomes, that is, the knowledge, skills, and attitudes that students are expected to master as a result of their educational experiences.

We use a variety of methods for collecting evidence that students are achieving intended learning outcomes, such as, written and oral questions, performance ratings, product reviews, journals, portfolios, and other self-report measures. These methods can collect evidence of student progress and achievement in a variety of teaching-learning environments. Gathering data and discussing information from multiple and diverse sources make it possible to know with confidence what students have learned. However, the learning assessment process is not complete until assessment results are used to improve students' educational experiences.

In this chapter, we emphasize the idea that in a culture that is cooperative, collaborative, and supportive, learning assessment is used to diagnose and promote learning. Teaching and learning are intertwined, and students and faculty learn together. We look, in detail, at the learning assessment process, describe selected assessment methods, and give examples of student learning assessment in representative programs. Finally, we identify key challenges to effective learning assessment, and point the way to addressing these challenges.

CHAPTER OBJECTIVES

This chapter is designed so that you can

- implement learning assessment processes
- create a plan to align assessment with intended learning outcomes and teaching-learning methods
- describe a variety of assessment methods that provide evidence of student learning

- use assessment results for the continuous improvement of learning experiences
- describe the key benefits and challenges to sound learning assessment

THE LEARNING ASSESSMENT PROCESS

In a traditional view, assessment is regarded as separate from teaching. Faculty believe that time devoted to assessment takes away from "teaching" time; students often regard assessment with dread and intimidation. In contrast, a CDIO approach views assessment as learning-centered, that is, an integral part of the teaching process, promoting better learning in a culture where students and faculty learn together. Table 7.1 is a comparison of teaching-centered assessment and learning-centered assessment, based on the work of Huba and Freed [1].

Assessment is learning-centered in that it is aligned with learning outcomes, uses multiple methods to gather evidence of achievement, and promotes learning in a supportive, collaborative environment. Assessment focuses on gathering evidence that students have developed proficiency in disciplinary knowledge, personal and interpersonal skills, and product, process, and system building skills. This student learning assessment is the focus of Standard 11.

> STANDARD 11 – LEARNING ASSESSMENT
>
> **Assessment of student learning in personal and interpersonal skills, and product, process, and system building skills, as well as in disciplinary knowledge.**

The process of assessing student learning has four key phases: the specification of learning outcomes, the alignment of assessment methods with

TABLE 7.1. **TEACHING-CENTERED VS. LEARNING-CENTERED ASSESSMENT**

• Teaching and assessing are separate	• Teaching and assessing are intertwined
• Assessment is used to monitor learning	• Assessment is used to promote and diagnose learning
• Emphasis in on right answers	• Emphasis is on generating better questions and learning from errors
• Desired learning is assessed indirectly through the use of objectively scored tests	• Desired learning is assessed directly through papers, projects, performances, portfolios, etc.
• Culture is competitive and individualistic	• Culture is cooperative, collaborative, and supportive
• Only students are viewed as learners	• Professor and students learn together

*From Mary E. Huba & Jann E. Freed. Learner-Centered Assessment On College Campuses: Shifting The Focus From Teaching To Learning.

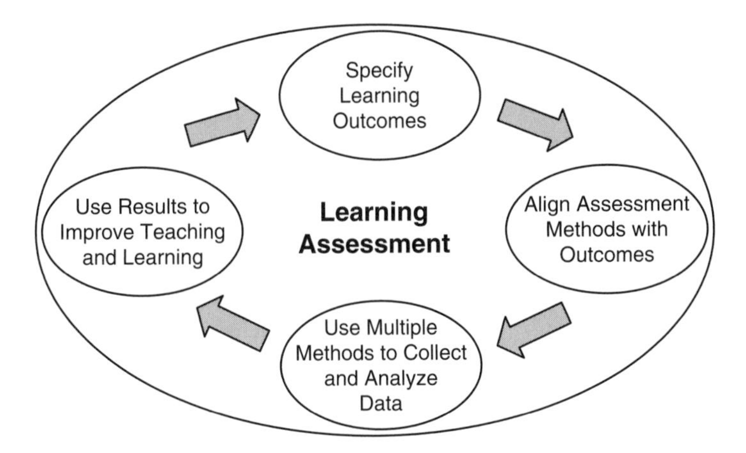

FIGURE 7.1. STUDENT LEARNING ASSESSMENT PROCESS

learning outcomes and teaching methods, the use of a variety of assessment methods to gather evidence of student learning, and the use of assessment results to improve teaching and learning. Figure 7.1 illustrates a process of learning assessment that can be implemented in any educational program. The unique characteristics of student learning assessment in CDIO programs are related to the nature of the learning outcomes, and their integration into the curriculum.

In most engineering programs, learning assessment focuses on disciplinary content. While this focus continues to be important in a CDIO approach, an equal emphasis needs to be placed on assessing the personal and interpersonal skills, and the product, process, and system building skills that are integrated into the curriculum. A single assessment method will not suffice to gather evidence of the broad range of learning outcomes.

Assessment of student learning begins with the specification of learning outcomes that students will achieve as a result of instruction and related learning experiences. Personal and interpersonal skills, product, process, and system building skills, and the disciplinary knowledge upon which they are based, comprise the overarching learning outcomes. In Chapter Three, we described the process of deriving learning outcomes from the CDIO Syllabus. (See Table 3.8) Once the learning outcomes are clearly stated, they are integrated into the curriculum, and sequenced for appropriate learning experiences. Chapters Four, Five, and Six describe this integration and sequencing in more detail. Just as different categories of learning outcomes require different teaching methods that produce different learning experiences—notably active and experiential learning approaches—they also require different assessment methods to ensure the reliability and validity of the assessment data. The next sections of this chapter address the second, third, and fourth phases of the learning assessment process.

ALIGNING ASSESSMENT METHODS WITH LEARNING OUTCOMES

Once the intended learning outcomes have been specified for a course, module, or other learning experience, they can be sorted into categories to facilitate the selection of appropriate assessment methods. For example, Table 7.2 is a general guide for selecting appropriate assessment methods aligned with specific categories of learning outcomes. It is adapted from the work of R. J. Stiggins, an educational assessment specialist who works with classroom teachers [2]. The first column identifies categories related to knowledge (3 levels), skills, and attitudes. Column headings are categories of assessment methods. Examples of assessment methods in the categories are explained in the next chapter section. For now, the table is given to emphasize the importance of selecting assessment methods that are appropriate for collecting evidence that students have achieved the specified learning outcomes.

The table suggests that conceptual understanding can be effectively assessed with written and oral questions. These questions might be included in examinations, interviews, or information interactions with students. Examples of learning outcomes in this category include:

- *Distinguish emissions from combustion characteristics (discipline-specific entry in CDIO Syllabus 1.3)*
- *Define a system, its behavior, and its elements (CDIO Syllabus 2.3)*

Problem solving and procedural knowledge can be assessed by asking students to find solutions to simple and complex situations with the use of oral questioning, written formats, or in reports and journals. Examples of learning outcomes addressing problem solving include:

- *Formulate solutions to problems using creativity and good decision making skills (CDIO Syllabus 2.1)*

TABLE 7.2. ALIGNMENT OF ASSESSMENT METHODS WITH LEARNING OUTCOMES

	Written and oral questions	Performance ratings	Product reviews	Journals and portfolios	Self-report instruments
Conceptual understanding	X				
Problem solving and procedural knowledge	X			X	
Knowledge creation and synthesis		X	X	X	
Skills and processes		X	X	X	X
Attitudes			X	X	X

- *Apply probabilistic and statistical models of events and sequences (CDIO Syllabus 2.1)*

Knowledge creation and synthesis learning outcomes, while more difficult to measure, can be assessed with some of the same methods as skills and processes. Examples include:

- *Conceive and design an engineering product to meet customer requirements (CDIO Syllabus 4.3 and 4.4)*
- *Appraise operations systems and recommend improvements (CDIO Syllabus 4.6)*

Learning outcomes that can be categorized as skills and processes are appropriately assessed with performance ratings, product reviews, journals, portfolios, and other self-report instruments. Examples of these outcomes include:

- *Determine the stress and deformation states of structures using the appropriate simulation tools (discipline-specific entry in CDIO Syllabus 1.3)*
- *Use appropriate nonverbal communications, for example, gestures, eye contact, poise (CDIO Syllabus 3.2)*

Finally, attitudes can be assessed with most self-report instruments, including journals and portfolios. The CDIO Syllabus specifies affective learning outcomes (attitudes) meant to be integrated into the curriculum. Examples include:

- *Recognize the ethical issues involved in using people in scientific experiments (CDIO Syllabus 2.2)*
- *Commit to a personal program of lifelong learning and professional development (CDIO Syllabus 2.4)*

The table is a *guideline* for matching assessment methods to learning outcomes; it does not prescribe exact matches. The choice of assessment methods often depends on a faculty member's experience with a method and available resources for data collection and analysis. The next section describes a variety of assessment methods and gives examples of their use in our programs.

METHODS FOR ASSESSING STUDENT LEARNING

The third phase of the student learning assessment process, as illustrated in Figure 7.1, is the use of multiple methods to collect and analyze data. Traditionally, assessment in engineering courses takes the form of written examinations, occurring usually at the end of the term. In contrast, student learning assessment in a CDIO approach uses a variety of methods to collect evidence of learning before, during, and after learning experiences to give a more comprehensive view of what changes have occurred in students' achievements and attitudes. Some assessment methods, when used during learning experiences, are effective as *teaching* methods, as well. For example, concept questions, described in Chapter 6 are effective both for learning new concepts and for giving instructors feedback on student learning. Evidence of student learning is gathered with written and oral questions, performance ratings, product reviews, journals, portfolios, and other self-report instruments. Criteria and standards of performance, incorporated into rating scales and rubrics, are used to assess the quality of student learning and achievement. We now look at a few of these data collection and assessment methods, and give examples from our programs.

Written and oral questions

Most engineering faculty are familiar with written examinations that include multiple-choice and other closed items, calculations, and open-ended questions. Faculty are encouraged to map their written examination questions to course learning outcomes, and to examine students' achievement in light of these outcomes. Written examinations continue to be effective and efficient means to assess students' conceptual understanding. A large number of students can be assessed in the same time period, and student achievement is documented. However, good questions are difficult to construct, and students' answers do not always reveal the causes of their errors or the sources of their misconceptions.

Oral questions, on the other hand, enable faculty to uncover students' misconceptions. Oral examinations require that students think on their feet and speak coherently. The use of oral exams, in conjunction with in-class concept questions, was described in Chapter Six in the example of an aerodynamics course at MIT. (See Box 6.1)

In both written and oral exams, faculty use concept questions to determine students' deeper level of understanding of disciplinary content. The use of concept questions, described in Chapter Six, is an example of a method that can is appropriate both for teaching and for assessment. Box 7.1 describes the use of concept questions to measure conceptual understanding in mechanics and mathematics at Chalmers University of Technology. The concept questions formed the second part of a longitudinal study that followed students over a three-year period.

Performance ratings

Many intended learning outcomes can be assessed by observing students in the performance of specific tasks, for example, oral communication and teamwork. In these situations, rating scales and rubrics facilitate the collection and analysis of assessment data. A rubric is a list of criteria that define the quality of a performance, process, or product, with a scale that reflects degrees of quality. In addition to their value to faculty, they convey expected performance to students. Since the same criteria are applied across the entire class, students feel that assessment is fair and more objective. Rubrics are effi-

Box 7.1. Excerpt of a Longitudinal Study of Students in Mechanical Engineering at Chalmers

The mechanical engineering program at Chalmers University of Technology conducted a study using the *Force Concept Inventory*, a well-known survey for testing students' understanding of basic physics concepts [1]. The FCI is more specifically designed to assess students' understanding of basic concepts in Newtonian physics, although it can be used for a variety of purposes and in many different contexts. The study compared the FCI survey completed in January 2002, with a second one completed in the Fall of 2002, when students were in their second year of the mechanical engineering program. Students were also asked conceptual questions about their understanding of physical concepts.

The following year, another study focused on mathematical modeling, that is, students' ability to use mathematics in applied situations. Students were given a well-known and frequently used mathematical modeling test, constructed by researchers in Australia, England, and Ireland, to collect evidence of growth in mathematical modeling competencies [2]. The modeling test was given twice, first in September 2003, when students had just started their third year of studies in the mechanical engineering program. The second survey was completed in February 2004, with the students in the second half of their third year.

Results of these longitudinal studies indicated that students in mechanical engineering showed increases in conceptual understanding of physics, mathematics, and mechanics from pre-test to post-test. Furthermore the *Force Concept Inventory* and the mathematical modeling tool proved useful in assessing students' achievement of learning outcomes related to disciplinary knowledge. The study showed differences in results by gender, but these differences could not be adequately explained in the context of the study. Further investigations were planned to determine which contextual factors affected the results.

[1] Hestenes, D., Wells, M., and Swackhammer, G., Force Concept Inventory, *The Physics Teacher*, vol. 30, 1992, pp. 141-151.

[2] Izard, J., Haines, C., Crouch, R., Houston, K., and Neill, M., Assessing the Impact of Teaching Mathematical Modeling: Some Implications, in Lamon, S. J., Parker, W. A., Houston, S. K., (Eds.), *Mathematical Modeling: A Way of Life*, ICTMA 11, Chichester, Harwood Publishing, 165-177.

– T. Lingefjärd, Chalmers University of Technology

cient means of recording observations and judgments, but their construction is time consuming and challenging.

Fortunately, examples of existing rubrics can be found in journal articles and conference proceedings of engineering education organizations, such as the American Society for Engineering Education (ASEE) and *Société Européenne pour la Formation des Ingénieurs* (SEFI). Table 7.3 is an example of a rubric that is used in some CDIO programs to rate the quality of students' technical briefings and oral presentations. Note that observers are asked to rate students' understanding of technical information, as well as their ability to present their ideas clearly and professionally. The sample rubric shown in Table 7.3 is an example of an analytic rubric. Here, the criteria are specified in some detail in the left-hand column. The evaluator makes a judgment about the quality of each criterion separately. In contrast, a holistic rubric starts with the gradations of quality, and describes in detail what, for example, a very good presentation would look like, with additional descriptions for each level of quality [2].

TABLE 7.3. SAMPLE RUBRIC TO ASSESS TECHNICAL BRIEFINGS AND ORAL PRESENTATIONS

	Poor	Fair	Good	Very Good	Comments
Presentation Quality Main objective of presentation is clearly stated. Presenter maintains good eye contact with the audience. Presenter uses voice effectively (volume, clarity, inflection). Presenter is poised and professional (appearance, posture, gestures). Transitions to the next presenter are smooth and effective.					
Technical Content Technical content is accurate and significant. Technical content shows sufficient development. Main points are emphasized and the relationship between ideas is clear. Ideas are supported with sufficient details and clear drawings. Graphics and demonstrations are effectively designed and used. Alternatives are presented with a rationale for those selected. Key issues are addressed. Questions are answered accurately and concisely.					

Overall Comments:

Product reviews

Similar rubrics can be developed and used to assess student products and projects. One of the key differentiating factors of a CDIO program is its emphasis on design-implement experiences. Students need to demonstrate their ability to conceive, design, implement, and operate products, processes, and systems. The assessment can be conducted by judging the demonstration (performance), or reviewing the physical product – be it an artifact, report, or computer drawing. Box 7.2 gives an example of a rubric that was designed and implemented at Queen's University Belfast to assess learning in a design project module.

A brief mention should be made here about *peer assessment*, that is, students assessing each other. Peer assessment most often occurs in the context of performance ratings and product reviews. For example, in the example in Table 7.3, students are given the rubric and asked to rate the oral presentations of members of their team or other class members. Of course, peer assessment does not have to be recorded with rubrics. In some programs, each student team is assigned the task of critiquing, both orally and in writing, the performance and products of at least one other team in the class.

Box 7.2. Project Design Review at Queen's University Belfast

The rating form embodies criteria designed to assess students' learning related to the project module outcomes. Project supervisors observe students and complete these forms.

The performance of the student during the course of the project will be assessed on the skills outlined in the table below. The supervisor is expected to rate the student's performance using the following scale:

Project Learning Outcomes	Unsatisfactory	Satisfactory	Good	Excellent
Communicated effectively in writing, verbally and through graphic media				
Managed time, resources, and priorities, and worked to given deadlines				
Used computers and information technology effectively				
Located and assembled information using various external resources				
Demonstrated generic problem-solving skills acquired during project				
Worked and learned independently				
Worked safely				
Communicated effectively with technicians and other support staff				

– R. Kenny, Queen's University Belfast

Journals and portfolios

Journals and portfolios provide records of students' individual and collaborative efforts in design-implement projects, and experimental research. They reveal students' critical thinking and reasoning skills, and record the steps students followed in an engineering process. These documents provide evidence of student achievement in situations where there may be no final tangible product. Moreover, journals help to distinguish individual contributions to group projects and activities. Although journals and portfolios take time to read and evaluate, they are most effective when students receive regular feedback.

Other self-report measures

Other self-report measures, such as inventories and questionnaires, help students to develop a sense of themselves as learners and future engineers. Asking students to reflect on their learning experiences helps them to see more clearly the connections among the concepts they have learned, as well as the applications of these concepts to new situations. When reflections are combined with portfolios that include samples of students' work, they serve as useful tools for assessing individual student achievement, and evaluating programs overall.

Box 7.3. The Use of a Reflective Portfolio at KTH

The reflective project was introduced in the vehicle engineering program at the Royal Institute of Technology (KTH) in Stockholm during the 2002-2003 academic year. Ten sophomore students were offered the opportunity to participate in four reflective portfolio sessions that met for three hours each. Five of the students participated through the end of the project two years later.

The specific outcomes of the reflective portfolio project were to enable students to:

- Reflect on learning methods and identify their own preferences for learning
- Increase learning efficiency through reflection
- Assess themselves and their peers
- Give and receive constructive criticism
- Plan their acquisition of relevant CDIO learning outcomes
- Identify and take advantage of opportunities for experience-based learning
- Develop habits of lifelong learning
- Take responsibility for career planning

Students were assigned short readings before each seminar and were required to bring a summary of their reflections to each project meeting. Discussion and comments for each student followed the sharing of reflections. The main idea was not to criticize, but to give other insights and experiences that might expand the ideas.

An electronic portfolio was established with three levels: a private level: a common group level; accessible with permission of the author; and, a public level, published on the internet and accessible to anyone. This portfolio contained exams, projects, writing, presentations, and so on. There were also personal narratives on what they represented in the students' learning.

The evaluation of the project showed that the students were very satisfied and found the reflective portfolio very illuminating. The dialogue and the reflection showed that they were not alone with their ideas, fears, and hopes, and that they shared many of those with other students. Project participants reported that they had new insights into their own learning, gained confidence in their ability to manage their studies, and developed a vision for their future studies and career. One student seemed to summarize students' perceptions of the project in the evaluation: "The project increased my awareness of my responsibility for my own learning."

– K. El Gaidi, KTH - Royal Institute of Technology

As mentioned before, students, themselves, can be involved in the assessment process by reviewing and commenting on the work of their peers, as well as their own work. These assessments may use rubrics similar to other performance rubrics, open-ended narratives, or reflective portfolios. Box 7.3 describes a reflective portfolio developed by students in the vehicle engineering program at the Royal Institute of Technology (KTH) in Stockholm for both peer and self-assessment.

USING RESULTS TO IMPROVE TEACHING AND LEARNING

The fourth, and perhaps the most important, step in the student learning assessment process—shown in Figure 7.1—is the use of assessment results to improve teaching and learning, and to help improve the program as a whole. This final step closes the assessment loop.

The elements of the learning assessment process are intended to help develop and demonstrate a *culture of quality*, as described by Massy [3]. He identifies a series of core quality principles that define the quality process for higher education.

- Define education quality in terms of student outcomes.
- Focus on the process of teaching, learning, and student assessment.
- Strive for coherence in curriculum, educational processes, and assessment.
- Work collaboratively to achieve mutual involvement and support.
- Base decisions on facts wherever possible.
- Identify and learn from best practice.
- Make continuous academic improvement a top priority.

Programs that regularly apply these principles in a process of continuous improvement demonstrate a culture of quality.

In the example of the project module assessment at Queen's University Belfast, described in Box 7.2 above, a number of instruments are used to identify the achievement of learning outcomes. The instruments are used to rate students' performance against the learning outcomes, thereby gathering direct evidence of student learning. The assessment instruments also provide a ready means of identifying where and how well the program's intended learning outcomes are being achieved. For example, if students are rated unsatisfactory by different examiners on any criteria on the rubric shown in the case study, then improvement efforts are targeted to remedy these deficiencies.

In addition to improving teaching and learning, assessment information may be gathered to satisfy institutional or external reviewers. For example, the criteria used in accreditation reviews of the Accreditation Board of Engineering and Technology state that a program should have in place:

- a curriculum that provides students opportunities to learn, practice and demonstrate student learning outcomes
- an assessment process that produces documented results that students have achieved the specified learning outcomes
- documented assessment processes with measurable student outcomes and feedback loops showing continuous program improvement [4]

Regardless of the requirements of external reviewing bodies, the most important use of assessment information is for the purpose of a program's own continuous improvement. The United States Naval Academy in Annapolis, Maryland, gives this example from its instruction for the annual reporting of assessment progress, "the fundamental purpose of Naval Academy academic assessment programs is to support continuous improvement of USNA academic programs and enhancement of midshipman learning"[5]. Such an ongoing process puts the periodic external review into its proper context and provides the most defensible use of the resources that are devoted to assessment. The use of student learning assessment data for program evaluation is addressed in more detail in Chapter Nine.

KEY BENEFITS AND CHALLENGES

Sound learning assessment methods contribute to student and program success in several ways:

- Gathering and discussing information from multiple and diverse sources makes it possible to know with confidence what students have learned.
- Teaching and assessment are intertwined, so that improving assessment methods also improves teaching and learning.
- There are appropriate assessment methods to measure student progress and achievement in a variety of teaching-learning environments.

In implementing sound learning assessment, several challenges remain:

- Shifting the traditional view of assessment to a more learning-centered approach is a challenge because engineering faculty tend to rely on the same assessment methods that were used in their own engineering studies, for example, problem sets and written exams. Support for enhancing faculty competence in assessment methods is described in Chapter Eight.
- Finding or creating reliable, valid, and appropriate learning assessment methods and tools matched to all learning outcomes can seem daunting, at first. The CDIO Initiative supports faculty and programs with the creation and implementation of new learning assessment methods. Collaborations encourages the sharing of assessment tools and results.
- Creating or adapting learning assessment methods that support deeper understanding of engineering concepts requires a serious commitment of faculty time. Formative assessment methods that take place within instructional units give faculty and students opportunities to monitor the development of knowledge, skills, and attitudes and reveal misconceptions. Summative assessment methods that take place at the end of instructional units give faculty and students opportunities to gain a broader and more cumulative perspective on the achievement of learning outcomes.
- The results of learning assessment, both formative and summative, are not always fully used. For maximum effect, results should be shared with faculty, students, and other instructional leaders to determine the extent to which learning outcomes have been achieved. The quality of these achievements becomes the motivation for continued excellence, or improvement, of future teaching-learning experiences.

SUMMARY

A program has implemented sound learning assessment when there is an explicit student learning assessment plan adopted by a majority of program faculty and other academic staff, and a variety of assessment methods matched appropriately to learning outcomes. Implementation is considered successful if a majority of faculty are using a variety of appropriate assessment methods, and using assessment results to determine student achievement to improve teaching-learning experiences in their courses. In short, a program has successfully implemented student learning assessment when there is evidence that all four phases of the learning assessment process, described in this chapter, are in place.

Finding or creating reliable, valid, and appropriate assessment methods and tools matched to all learning outcomes remains a challenge. The CDIO Initiative offers opportunities for faculty and assessment specialists to develop new tools and share their experiences across a variety of engineering education contexts. Collaborators are building an inventory of assessment approaches, for example oral exams, performance rubrics, portfolios, self-report instruments, as seen in the examples in this chapter.

This chapter examined alternatives for planning sound learning assessment and suggested ways to overcome key challenges to its implementation. Establishing a culture in which assessment promotes learning requires a shift in perspective from a teaching-centered to a learning-centered approach, and a commitment to use assessment results to improve curriculum, teaching methods, and the overall learning environment. Issues related to expanding faculty competence in the use of student learning assessment methods, as well as teaching-learning methods, are addressed in the next chapter.

DISCUSSION QUESTIONS

1. What types of data or evidence do you rely on most heavily in your decisions about engineering courses and programs?
2. What assessment methods can you introduce or improve in your courses?
3. How do you use learning assessment results to improve curriculum, teaching and learning, student and instructor satisfaction, and learning spaces in your programs?
4. How would you begin to address the key challenges to assessment posed in this chapter?

References

[1] Huba, M. E., and Freed, J. E., *Learning-Centered Assessment on College Campuses*, Allyn and Bacon, Boston, MA, 2000.

[2] Stiggins, R. J., *Student-Centered Classroom Assessment*, 2d ed., Merrill, Upper Saddle River, NJ, 1997.

[3] Massy, W. F., *Honoring the Trust: Quality and Cost Containment in Higher Education*, Anker Publishing, Bolton, MA, 2003.

[4] Accreditation Board of Engineering and Technology, *Evaluative Criteria 2000*. Available at http://www.abet.org/criteria.html

[5] The United States Naval Academy, Academic Dean and Provost, Instruction 5400.1 (internal document)

CHAPTER EIGHT ADAPTING AND IMPLEMENTING A CDIO APPROACH

WITH D. BODEN

INTRODUCTION

Adapting and implementing a CDIO approach can potentially be of great value to educational programs and the students they serve. However, that means change—an inherently challenging endeavor, especially at a university. Program leaders are more likely to succeed in this change process if faculty are equipped with an understanding of how to bring about change, and provided relevant guidance and resources. This chapter discusses the implementation of a CDIO approach in terms of three change processes: cultural and organizational change, faculty development and support, and program change.

The transformation to a CDIO program will touch all faculty members in the program, and it will influence the context and the organization of the education program. To succeed, faculty need to view this as an instance of cultural and organizational change, and to make use of the lessons learned that facilitate such change processes. The first section of the chapter reviews these lessons and applies them to the university context.

Implementation places new demands on our faculty and teaching staff. We cannot expect them to acquire new skills without the resources to enhance their own competence. In the second section of the chapter, CDIO Standards 9 and 10 will be introduced. Standards 9 and 10 deal with the issue of faculty competence in professional skills and teaching. This section will also discuss approaches that enhance the competence of current faculty and build a stronger faculty in the future.

The CDIO Initiative supports the change process and the enhancement of faculty competence by developing resources and frameworks. The third chapter section presents a roadmap for program change that represents adaptation and implementation as an engineering design process. This third section outlines examples of supporting resources that are currently available.

CHAPTER OBJECTIVES

This chapter is designed so that you can

- recognize key success factors that influence change in an organization
- view the development of a CDIO program as an example of cultural change
- plan activities that enhance faculty competence in personal and interpersonal skills, and product, process, and system building skills
- plan activities that enhance faculty competence in teaching, learning, and assessment methods
- describe approaches and locate resources that facilitate the adoption and implementation of a CDIO approach in engineering programs

DEVELOPMENT OF A CDIO PROGRAM AS AN EXAMPLE OF CULTURAL AND ORGANIZATIONAL CHANGE

As described in Chapter Two, the dominant paradigm for engineering education today is one in which the *content* is disciplinary and based on engineering science. The unintended consequence of the transformation in the last century to this paradigm is that the *context* of the education also became based on engineering science and research. Some degree of cultural change is required to transform a program to the desired vision, one which better integrates the engineering science disciplines and sets them in the context of conceiving-designing-implementing-operating products, processes, and systems.

Fortunately, there is broad understanding of the factors that support successful cultural change in organizations. Adapting these change factors to the university environment facilitates the transition to a CDIO approach. We begin our discussion of change by re-examining the two central questions that were posed in Chapter Two:

- *What is the full set of knowledge, skills, and attitudes that graduating engineers should possess as they leave the university, and at what level of proficiency?*
- *How can we do better at ensuring that students learn these skills?*

The approach to answering the first question was largely answered in Chapter Three with the discussion of the CDIO Syllabus. Chapters Four to Seven presented approaches to answering the second question. But how can we convince our colleagues of the need to "do better at ensuring that students learn these skills?"

The second question is deliberately posed in the language of continuous process improvement. Of course we can do better! We can benchmark our peers and learn from best practice. We can better acquaint ourselves with learning theory and apply it appropriately. We can listen to our internal and external stakeholders. We can learn to better apply technology. We can do better. Our students ask us to, our industry stakeholders prod us to, our government and professional regulators encourage us to. It is not a matter of fixing something that is broken, but of improving something that is vital to our future—technological education.

However, adapting and implementing a CDIO approach will be somewhat of a challenge at most universities, and for good reason—it requires *change*. The words *change* and *university* do not easily sit in the same sentence. There are two perspectives that must be understood before beginning the process of change at a university:

Perspective 1: Universities are, by design, resistant to change as organizations.
Perspective 2: Notwithstanding Observation 1, universities *can* be changed by appropriate application of best practice in leading organizational change.

Universities are resistant to change as a matter of organizational design and tradition. In Europe, universities emerged in the Middle Ages from cathedrals and monastic organizations, designed for stability and contemplation in an era of societal chaos. Universities adopted the concept of tenure—the right to contradict established views. At best a license to be bold and innovative, tenure makes difficult the concept of organizational alignment. Despite the fact that they appear hierarchic (departments, deans, provost, others), universities are actually very flat organizations in which lines of authority are fuzzy, weak or nonexistent. Historically, universities tended to be introspective, relying on self-reflection and debate to drive change.

As a result, universities are stable, long-enduring institutions. In Europe, of the more than 25 institutions that have operated continuously since the Reformation, all but four are universities. In the United States, there were already nine universities as early as 1770 [1]. More than one university in Europe and the United States shared the outlook that "a little change is good, no change is better."

Yet change can be effected at universities. In the 1880s, Eliot changed Harvard from a colonial college to a modern university [1]. Vannevar Bush created a new frontier in *Science The Endless Frontier*, fundamentally altering the view of research at universities in the United States [2]. As we pointed out in Chapter Two, the engineering science revolution of the latter half of the 20th century fundamentally changed the approach to engineering education worldwide. This evidence suggests that change, even major transformational change, is possible in universities. But leaders are more likely to be effective in bringing about change if they are equipped with an understanding of how to lead organizational transformations.

Key success factors that promote cultural change

What forces can be brought to bear on a university to catalyze change? Our second observation is that, much to our surprise, change is precipitated at universities in much the same way as in most organizations. In short, there must be strong leadership, vision, plans and action, resources and incentives. We will discuss twelve key success factors that can be used to guide change at a university. These factors fall broadly into three categories—getting started, building momentum, and institutionalizing change:

- Getting off to the right start
  - Understanding the need for change
  - Leadership from the top
  - Creating a vision
  - Support of early adopters
  - Early successes
- Building the momentum in the core activities of change
  - Moving off assumptions
  - Including students as agents of change
  - Involvement and ownership
  - Adequate resources
- Institutionalizing change
  - Faculty recognition and incentives
  - Faculty learning culture
  - Student expectations and academic requirements

Each of these success factors will be discussed in turn below, with a view towards the general issue, and its application to adapting and implementing a CDIO approach in a university program.

The first phase of change—getting off to the right start

There are five key success factors that help the change process in its initial stages: understanding the need for change, leadership from the top, creating a vision, support of early adopters, and early successes.

Understanding the need for change. There must be a stimulus and motivation for change. The stronger and more clearly understood this need is, and the more urgent, the more willing the organization will be to change. Crises and external threats are classically strong motivators, but universities are usually somewhat insulated from such influences. However, it is vital that in this change process, the team involved understands the need, and is committed to addressing it. Since a university acts as a collective of faculty and staff, it is important to articulate this need for change in such a way that groups, as well as individuals understand it.

As this is an educational change, it is best to focus on the needs of the students, those who benefit from the education. What are their needs? According to whom? This issue is sometimes cast as dissatisfaction with the current situation. Alternatively, it can be cast in the terminology of continuous process improvement—*can we do better at meeting the needs of the students?*

The stimulants we have found successful at precipitating an examination of needs rely, in large part, on external references. The inputs from industry, discussed in Chapters Two and Three, are examples. Guidance from alumni of the program is valuable, as are the inputs from external review committees and external members of program boards. It is also useful to cite the opinions of thought leaders and "authorities." If taken constructively, a national accreditation process can also be an external stimulus for change, as it was at the United States Naval Academy. (See Box 8.1).

Box 8.1. The Adoption and Assessment of a CDIO Approach at the U. S. Naval Academy

The Department of Aerospace Engineering at the United States Naval Academy joined the CDIO Initiative in July 2003. The CDIO Initiative provides us with the framework and tools necessary to make and assess changes in our program. The Naval Academy produces officers who serve in the United States Navy and Marine Corps. Therefore, the goals and outcomes of all the academic programs, including the Aerospace Engineering program, support the Naval Academy mission under the leadership of the Academic Dean and Provost. The institution has developed a set of strategic educational outcomes that describe the results it wishes to produce in the graduates. The mission of the Aerospace Engineering Department must follow from the mission of the Naval Academy, while at the same time emphasize the role of the aerospace engineering major. Our mission is to:

> *Provide the Navy and Marine Corps with engineering graduates capable of growing to fill engineering, management, and leadership roles in the Navy, government, and industry; maturing their fascination with Air and Space systems.*

Our departmental vision follows our mission:

> *Mission fulfillment requires a program wherein Midshipmen Conceive-Design-Implement-Operate complex mission-effective aerospace systems in a modern team-based environment.*

Both our departmental mission and vision are a direct result of our participation in the CDIO Initiative.

Initially, our primary interest was the approach to program assessment that is tied to the CDIO Syllabus. We felt that this approach would be of great assistance in meeting the new accreditation standards set forth by the Accreditation Board of Engineering and Technology (ABET). However, as we learned more about the CDIO approach, we were convinced that it was right for us for many reasons beyond our initial interest in the assessment process. The primary reasons for adopting a CDIO approach in the Aerospace Engineering program at the Naval Academy were:

- Our desire to go beyond "paper designs" in capstone design courses
- The strong focus at the Naval Academy on operations—our graduates become operators of systems
- To have a structure to make necessary changes in our program
- To benefit from lessons learned from the four founding universities that would help guide our design and implementation of a renewed Aerospace Engineering program

Once we decided to adopt the CDIO Syllabus, our next question was, *how do we gain support from the administration, the program leaders, and the faculty?* Once we completed the CDIO Syllabus and looked at our existing program, it was clear that we valued topics in the Syllabus, but we were not teaching the topics. This discrepancy provided the motivation for change and made the job of convincing our faculty to adopt a CDIO approach. The survey of our key stakeholders further solidified the need for change and the advantages of the CDIO Syllabus and approach.

– D. Boden, United States Naval Academy

Another valuable means of external reference is benchmarking. In some systems, universities share data; in others, the government publishes data on performance. Top universities are constantly benchmarking themselves against peers, through both formal and informal means.

Pressure from above is another way to stimulate commitment. If a university president or school dean mandates examination, strategic planning, or program re-assessment, it is a wonderful opportunity to catalyze action at a department or program level.

Barring external stimuli, it is possible to create an internal urgency by framing the issue in the terminology of continuous process improvement. It is important to avoid the formulation that something needs to be fixed, and ask instead *"how we can make it better"*.

New resources can be a catalyst for change. New faculty positions, new building funds, new equipment funds, new calls for proposals from government agencies can all be used to bring focus to change, especially if the resources are to be awarded competitively.

Finally, large-scale social shifts can provide the context in which change is more possible. In the United States, the competition with the Soviets, made clear by the launch of Sputnik, was a catalyst for massive educational change. In Europe, the recent Bologna Accord has created great fluidity in higher education thinking [3].

Leadership from the top. Leaders are in the best position to change a culture. The commitment of the leader, and his or her active participation, is vital. In the case of a university department or program, the department chair, or program head, must lead the change process. Delegation to a committee, or junior member of the team, will almost certainly produce weaker results.

The formal leader must be supported by a strong *inner* team of recognized individuals in the program. In order to change an organization, thought leaders must visibly demonstrate their interest and participation. This inner team can be made up of senior and junior faculty members who are effective as innovators. In addition to providing visible support, the team can serve as a sounding board, brainstorming, and planning group. It is not advisable to make this group exclusive, for that might create an "us vs. them" sense with the extended population. Rather, the inner group needs to be porous and inclusive.

It is also desirable to have the visible support of people in the organization who are at one or two levels above the change leaders. Deans, provosts, rectors, and vice-chancellors provide resources and organizational authority. They often seek change in the organization, but are too remote from the faculty to lead effectively at individual department or program levels. Therefore, they are often pleased and supportive of proposals for change from departments, programs, or schools.

Creating a vision. In promoting change, it is most helpful if the leader, sometimes aided by a small group, quickly communicates a vision of how the urgent needs will be addressed. This vision should be easy to communicate, and become the organizing theme of the work. It may evolve into a full-blown strategy for change.

In consensus organizations, such as university departments or programs, tension often arises on this point. On one hand, if the inner group arrives at the vision too quickly, there will be a sense that it was imposed, losing the values of broad-based ownership that are important for long-term acceptance. On the other hand, delaying too long before reaching consensus on a vision leads to organizational confusion and loss of any potential acceptance. The leader must strike this balance appropriately.

We have adopted an explicit vision, conveyed in Chapter Two—that engineering education should be set in the context of conceiving-designing-implementing-operating products, processes, and systems, and that the education is best done by organizing around the disciplines with design-implement experiences interwoven. We have formally incorporated this vision into Standard 1. In adapting a CDIO approach to a particular program, it may be useful to start with this premise, and build a vision. Alternatively, it is also possible to start with an organization-specific version, for example, *the whole engineer*, and then build on the similarity with the CDIO vision.

Support of early adopters. In any population, on any issue, there are those who are more inclined to try new approaches, those who will wait a bit, and those who will tend to resist change. The first group is generally referred to as *early adopters*. These individuals can be very important agents of change. Early adopters should be included in the change process as quickly as possible. To the extent possible, they should be given resources to develop pilots or experiments. If successful, these efforts should be celebrated. In this way, momentum will be begin, and not-so-early adopters among their peers will become curious and engaged.

The program leader should identify early adopters at the beginning of the change process, and encourage them to join the effort. Academic departments and programs are often small enough that known attitudes and performance make it easy to identify the early adopters. Students are a good source of information. They can often recognize the dedicated educators. Steps as simple as inviting an outside speaker on education or calling an optional meeting and seeing who shows up will often identify these individuals. In brief, it pays to identify early adopters, engage them, support them and celebrate their successes.

Most universities have organizations dedicated to educational research, development, and support, staffed by professionals in education and pedagogy. Another form of support that can be given to early adopters is collaboration with staff from such an educational support service organization. These groups are often enthusiastic about the prospect of participating in larger-scale reform, and are, themselves, among the early adopters.

Early successes. It is important to achieve some visible successes early, in order to attract interest and stimulate the effort for change. Often in the reform of academic programs, there is a long planning process, sometimes stretching over years. Educational reform is more likely to succeed if there is a spiral process, in which early goals are identified, early pilots run, the results reflected upon, and new goals set. From that point the process continues. Positive outcomes of these early pilots often attract the support and interest of others. These early successes are often developed by teams of early adopters.

In starting the change process, the leader should explicitly plan on developing some successes quickly, in the first term, or sooner if possible. Ideally, these early successes would have high visibility and wide impact. They should be recognizable as making the education better or the job of the teaching staff more productive. Examples that we have developed include modifications of:

- A first-year course to include a basic design-implement experience
- An upper-level course to include more comprehensive, yet low-cost, design-implement experiences
- An appropriate meeting room or flexible classroom to create a design-implement workspace that supports hands-on and social learning

As a program transitions from this first phase of "getting off to the right start", it is important to reflect on its progress and accomplishments. Box 8.2 summarizes observations from the program leader of the Mechanical and Materials Engineering Department at Queen's University in Kingston, Ontario (Canada) about this transitional point in the adaptation and implementation of a CDIO approach.

The second phase of change—building momentum in the core activities of change

There are four key success factors of cultural change that build momentum: moving off assumptions, including students as agents of change, involvement and ownership, and adequate resources.

Moving off assumptions. Once the change process is underway, the leader needs to get the team to move off their traditional assumptions of what and how things should be done. A successful change process requires willingness to think outside of the box, and to try new things. Despite academic commitment to research, scholarship and innovation, organizational willingness to change is not a strength of most universities.

There are several approaches that can be used to stimulate flexibility. A powerful one is an appeal to professionalism. Faculty members are often dedicated and distinguished professionals in their respective fields of engineering. If you can appeal to their professionalism as engineers, and transfer that sense of professionalism to education reform, you will harness an important force. This appeal can be accomplished by posing the change process as an engineering design problem. This immediately raises questions such as: *What are the requirements? What technology is available? How can we create prototypes?* Such questions engage faculty in a new way. An expanded discussion of this point is found in the chapter section that describes the third phase of the change process. One can imagine applying this approach in other domains as

Box 8.2. The Adoption of a CDIO Approach at Queen's University (Canada)

The Department of Mechanical and Materials Engineering in the Faculty of Applied Science at Queen's University is one of the larger departments of its kind in Canada. Its program includes many technical electives from which students can choose, making it attractive to students because they can find employment in many different areas. The departmental research strengths are in energy systems, biomechanical engineering, manufacturing, and materials.

In late 2002, Ed Crawley from MIT introduced the CDIO approach at Queen's University to faculty and others interested in engineering education. A few members of the department then participated in the collaborator meetings at the Massachusetts Institute of Technology and the Technical University of Denmark to learn more about the CDIO Initiative. Their reports were convincing enough to get unanimous departmental support for joining the Initiative in December 2003. The work of the Initiative that had been done already in 2003 was seen as a boost to ongoing activities at Queen's that were supported by the Dean of the Faculty. Courses and ideas similar to those advocated by a CDIO approach already existed to a limited extent in the departmental program. Furthermore, the recent completion of the *Integrated Learning Centre* (ILC) provided an ideal facility to support a CDIO curriculum.

Feedback from an annual industrial review board, student evaluations, and faculty initiatives all pointed in the same direction. More emphasis was placed on *conceive, design, implement and operate* exercises, communication, teamwork and other professional skills, without, of course, sacrificing the teaching and learning of basic engineering knowledge and skills. It was clear that the CDIO Initiative had gone further by gathering feedback on the engineering curriculum from students, faculty, alumni and industry in the United States and Sweden. It had also developed a syllabus containing the necessary elements of an engineering curriculum, together with many other supporting initiatives from which it was obvious that Queen's could benefit. At the same time it was felt that Queen's, through its own initiatives, could also contribute.

The following are some of the activities that have been undertaken at Queen's since joining the CDIO Initiative. Without the impetus provided by the collaboration, these activities would not have been done at all, or would have been undertaken only to a very limited extent.

- An alumni survey with over 400 respondents
- Benchmarking of the program against the CDIO Standards
- Adding "C-D" in one capstone course and "I-O" in another
- Improving the way we provide technical communication training

There is still a lot more work that needs to be done to improve the curriculum of Mechanical and Materials Engineering. It is clear that continued participation in the CDIO Initiative will allow us to discuss and learn more from other collaborators, thus allowing the Department to develop an improved program.

– U. Wyss, Queen's University (Canada)

well. If the reform of medical education, the leader might ask: *How can we diagnose and heal our patient—the ailing medical educational system?*

Evidence is an important way to move peoples' opinions. We tend to underutilize evidence-based approaches in universities. *Just exactly what are thought leaders outside your institution saying? What is the status of other initiatives at peer universities? What are other departments or programs in your university doing? What resources are available?* Compiling briefing books of such evidence provides an opportunity for evidence-based policy change.

One way to engage the faculty in reviewing this evidence is to conduct an Oxford- or Cambridge-style formal debate. In this format, the faculty member does not necessarily argue his or her own personal opinion, but a preassigned position, either for or against the issue, based on the evidence or other accumulated facts that are presented. Clever positioning of the individuals involved will often cause a person to be arguing a position counter to his or her personal opinion. If faculty members know they are playing roles, most will try hard to be persuasive, and perhaps, even persuade themselves.

Another common technique to make a group feel comfortable with change is called *stretch and relax*. This requires first *stretching*—considering change that is credible but slightly extreme; then, allowing the team to *relax* back to a position that is less extreme, but still forward of the status quo. This is the essence of the art of political compromise.

Including students as agents of change. Another force that can either promote or delay change is student opinion. We have found that it is important to include students in a substantive way in the educational change process. On the positive side, students can be powerful agents for change in their own right. They tend to know what works well, and what does not—who teaches well and who does not. They can be a valuable source of information in planning for change, and in providing feedback after changes have been piloted. This role is enhanced if we explain to students the motivation behind the change, and the direction in which it is going. Once students experience program change, they often put pressure on the not-so-early adopters to improve, as well.

On the negative side, students can be uncomfortable with change in the same way that faculty might be. Students are particularly threatened by change in their personal futures. An approach to overcoming this fear is to create rolling change, starting in the early years of the program. In this way, the more senior students can advise students in subsequent years about change that will occur in the program.

It is important to include students, in varying degrees, in the decision making process at the university. Students can act as important agents of change. Engage formal student groups, and invite individual thoughtful students to participate in discussions. Appeal to students' professionalism, as well. Consider giving them a major role in the change process.

Involvement and ownership. It will eventually be necessary to involve all of the members of the team in the change process. Our experience is that it is better to do this early. Academic programs are *owned* and implemented by a wide variety of faculty members, some of whom may not be formal members of the department or program. For the reforms to take hold and be executed in individual courses, all of the participating faculty must be at least satisfied with, if not enthusiastic about, the change.

The initial inclination in launching a change effort in a department or program is to form a committee of early adopters to plan the change. While this has its advantages, it may appear exclusive. It requires a two-step process of first working with this group, and then having this group influence the larger group. A preferable, but more awkward approach, is to engage the entire group as a committee of the whole, or at least invite all to participate—an Athenian democracy. If the leader must work with a smaller group, it should be representative of the various interests of the larger group, so that you have allies from all interest groups in the larger selling process. It is sometimes a valuable device to give an important task to a known skeptic. If that person becomes convinced, he or she will bring along many others.

In gaining involvement and ownership, it is important to allow participants time away to reflect and debate. Traditional approaches to this are off-site meetings, away days or retreats. It is important to plan these times carefully, laying out agendas that actively engage the group, and selecting effective discussion facilitators.

Adequate resources. Sustainable change must be accompanied by adequate resources. While it is unlikely that there will be significant new resources available to the program in the long-term steady state, it is also true that educational change cannot be achieved on the margin. The transformation will require time and interim support, which must be made available to the participating instructors and members of the staff.

These transitional resources will normally take the form of a term with release from teaching, and extra teaching support, such as assistants and other help. In some universities, it is possible to obtain support for academic projects beyond the academic year. In a culture that values engineering science and research, it is important to make the statement that time dedicated to reforming education should be part of the educational pool of resources, not drawn from the time allotted to research.

We have had two main goals with regard to resources. First, we have tried to create a collaborative effort, which has produced the open-source resources described in a later section of this chapter. This combination of approach and shared resources minimizes the additional time and energy that must be expended by a program in the transition. Second, our objective is that, in the steady state, a program would need no new resources; instead, existing resources would be retasked.

The third phase of change—institutionalizing change

Three key success factors facilitate the institutionalization of change: faculty recognition and incentives, a culture of faculty learning, and student expectations and academic requirements.

Faculty recognition and incentives. A maxim of sustainable change is that incentives must be aligned with change. In any organization, you get the behavior that is rewarded. If education is important, the leader of the program or department must create both the perception and reality of incentives and recognition that reward education, in general, and educational reform, in particular.

The leader must be supported by those at higher levels in the university in this effort to reward sustainable change.

Many universities have recognition programs for faculty in the form of teaching awards. These awards can be combined with the presentation of special status to those who are known for particularly good teaching, such as *chairs for teaching*, or other university-wide recognition. This group of honored faculty can convene and discuss educational innovation throughout the university and act as an academy for education within the university. Occasionally, recognition for faculty contributions to education comes from national academies of engineering or other honorific bodies.

It is vital that the formal review process recognize and reward educational contributions. This means that in annual review cycles, submissions of educational contributions should be reviewed. More importantly, in hiring, promotion, and tenure decisions, weight should be placed on these efforts. In particular, scholarly publications in education should be valued as highly as those in other scholarly research publications.

Faculty learning culture. All universities place great emphasis on learning, particularly on the broad learning of students, and of the faculty in their professional disciplines. Ironically, many university faculties do not also embrace a culture that broad life-long learning by the faculty is important outside of their discipline. Leaders of change must create the expectation, and set the standard, that life-long learning of instructors is important, not only in their professions, but also in education and the teaching of professional skills.

Movement in this dimension is not difficult, and often begins with simply making this observation and setting this expectation. Action can include granting faculty leaves and sabbaticals for professional engineering activities or educationally related activities, in addition to research sabbaticals. A department or program can begin to circulate important writings on these topics and discuss these topics at meetings, much like a research group reads the current literature.

Faculty members can be asked to develop their own professional development plan, perhaps as a part of an annual review. They might be challenged to define what and how they are going to learn in the next year. If the leader sees a pattern in the learning needs of the faculty, he or she can create learning opportunities at the university by bringing in outside experts, providing short courses for the faculty, or making connections to other university groups, including the university teaching and learning center.

Student expectations and academic requirements. Students are the immediate customers and beneficiaries of the educational service we provide, and as in any change process in a customer service organization, their expectations must be carefully considered. This takes two forms: their informal expectations and the formal academic requirements.

We have observed that, like all first impressions, a program has one chance to set expectations for learning and behavior in its students. At a university, this is the first day of the first year or, the first day students enter the program. At this first opportunity, the goals of the education should be explicitly explained. More importantly, the expected norm of student involvement in learning must be immediately established. If we are to expect more active learning, with more student responsibility for their own learning, we should establish this pattern on the first day of instruction, and consistently throughout all of the classes or modules the student encounters.

It is also desirable to institutionalize the learning outcomes in formal descriptions of the academic program. This can be done at the course or module level, in the learning outcomes and objectives, and, at a higher level, in the overview or description of the course of studies or program.

Change at a university as an instance of organizational change

The key factors that facilitate change at a university are not dissimilar from frameworks that have been developed to help other types of organizations succeed in change. For example, there is a general consensus that the start of the process is critical, and that the force of urgency and understanding of need, vision, and first steps must be coordinated to overcome resistance to change. In fact, there is even a *formula* based on this observation, credited to Beckhard, Harris and Gleicher [4].

$$D \times V \times F > R$$

The formula is read as: dissatisfaction, D, (a measure of the understanding of the true need and the opportunity to improve) times vision, V, times first steps, F, must be greater than the resistance to change, R. Other frameworks, such as the one developed by Kotter [5] outline steps generally applicable to organizational change (Table 8.1). These emphasize the beginning of the effort and bear close resemblance to some of the twelve success factors for universities, described in this chapter.

This comparison with the broader literature and experience with organizational change should give us reassurance that with commitment, attention to process, and sensitivity to the unique characteristics of the university environment, we should be able to effect the changes necessary to implement a CDIO approach. Box 8.3 describes an application of a change process model, called AWAKEN, at the University of Liverpool.

FACULTY DEVELOPMENT AND SUPPORT

As is evident in discussion of the key success factors above, faculty involvement and enthusiasm greatly facilitate the implementation of a CDIO approach in engineering programs. Faculty are asked to be innovators—they are asked to adapt their teaching style to one that is more student-centered,

TABLE 8.1. **KOTTER'S EIGHT STAGES OF THE CHANGE PROCESS [5].**

1	**Establishing a sense of urgency**
	Examining market and competitive realities
	Identifying and discussing crises, potential crises, or major opportunities
2	**Forming a powerful guiding coalition**
	Assembling a group with enough power to lead the change effort
	Encouraging the group to work together as a team
3	**Creating a vision**
	Creating a vision to help direct the change effort
	Developing strategies for achieving that vision
4	**Communicating the vision**
	Using every vehicle possible to communicate the new vision and strategies
	Teaching new behaviors by the example of the guiding coalition
5	**Empowering others to act on the vision**
	Getting rid of obstacles to change
	Changing systems or structures that seriously undermine the vision
	Encouraging risk taking and nontraditional ideas, activities, and actions
6	**Planning for and creating short-term wins**
	Planning for visible performance improvements
	Creating those improvements
	Recognizing and rewarding employees involved in the improvements
7	**Consolidating improvements and producing still more change**
	Using increased credibility to change systems, structures, and policies that do not fit the vision
	Hiring, promoting, and developing employees who can implement the vision
	Reinvigorating the process with new projects, themes, and change agents
8	**Institutionalizing new approaches**
	Articulating the connections between the new behaviors and corporate success
	Developing the means to ensure leadership development and succession

*Reprinted with the kind permission of Harvard Business School Publishing.

BOX 8.3. THE AWAKEN MODEL

The AWAKEN model is one of the techniques that underpin implementation of the CDIO approach at The University of Liverpool. Embracing Bloom's Taxonomy [1] of the cognitive, affective, and psychomotor domains, AWAKEN strives to provide a humanistic, democratic and science-based methodological approach to sustainable adaptation and agility within the education process. What results is an open, flexible, and change-tolerant learning organization.

AWAKEN's six-step approach captures the essence of the *Balanced Scorecard* [2], harnessing and clarifying the overarching vision and strategy, and formulating an achievable and agreed-upon action plan for implementation that consistently and efficiently meets stakeholder expectations. AWAKEN thus becomes both a structure and procedure through which all *education lifecycle* perspectives are captured, understood, and managed, such that a sustainable education environment results. AWAKEN's quantitative and qualitative processes provide a systemized, yet flexible, implementation process, making it applicable at module, program, inter- and intra-organizational levels. AWAKEN is currently being deployed in aerospace and electronics organizations in the United Kingdom.

(*Continued*)

Box 8.3. The AWAKEN Model—Cont'd

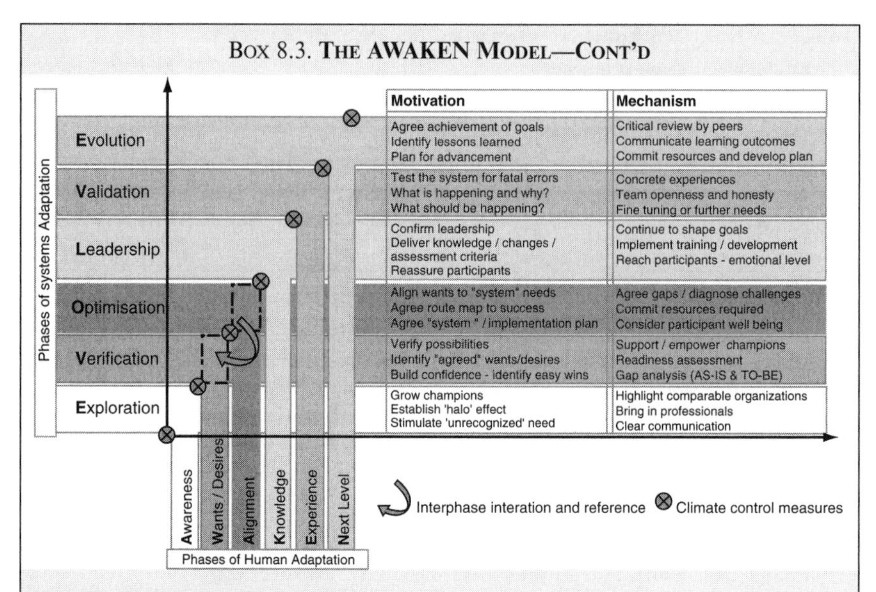

The AWAKEN Model: Baseline Motivations and Mechanisms

AWAKEN in the context of the CDIO approach aligns the local strategic learning and teaching objectives of the Department of Engineering with the needs of the evolving local school-university-industry system (the process), the needs of education specialists within the system (faculty members), system users (students), and external stakeholders of the system (employers), against a backdrop of the CDIO Syllabus. The fundamental challenges in implementation are related to the identification of the markers of successful implementation and assessment. In particular, the key questions focus on the effective assessment and self assessment criteria required to evaluate knowledge, skills, and attitudes. One mechanism being considered to assist in the deployment of self-assessment is the Individual Competency Record (ICR). ICR aims to capture key Learning Outcomes (LO) against the generic CDIO phases of systems development and integration. Each student will make journal entries and notes related to particular LOs, forming a portfolio approach to self-directed learning and review. Tutors will appraise progress against LOs on a weekly or fortnightly basis.

ICR

Module Title	What's it made of?														
MATS109	All			Plan		Design		Source		Make		Sell		Service	
Learning Outcomes	Quality	Project/Program Management	Change/Release Management	Concept	Planning	Design	Engineering	Advanced Sourcing	Volume Sourcing	Manufacturing Planning	Production	Market Rollout	Sell	Service	Retirement
Range of materials	⊙														
Classification of materials			⊙												
Material selection															
Use of CES															
Group working					⊙	⊙									
Problem solving						⊙		⊙					⊙		

In Mech210 we analysed the material properties of a failed Maserati suspension component using metallographic techniques and found ......

Sample Individual Competency Record

One of the hypothesized measures of successful implementation is the concept that there will exist an identifiable confidence transition point (CTP) that will be evidenced by system *ownership* of the new CDIO approach by participants. This hypothesis is yet to be shown.

AWAKEN and the CDIO Confidence Transition Point (CTP)

One of the greatest obstacles to any implementation is fear and past practice. Change, or rather adaptation, on this scale brings about much fear and uncertainty, which left unmanaged during the process, may prove disruptive. It is known that behavior is largely driven by expectation derived from historical experiences rather than perceived future scenarios. To counter any negative forces, the AWAKEN-CDIO implementation process deploys *Change Conversations* (CC), as used by Jack Welch during his reign at General Electric [3]. A CC is a regular newsletter to inform all stakeholders of progress toward the goal, sowing the seed that a CDIO approach is a wholly positive strategy, and is organizationally important. This serves to reinforce the vision that taking action will yield long-term positive culture change within the environment. Through an industry-informed CDIO Syllabus of engineering knowledge imbued with rich visceral sensory engagements of real life, the CDIO approach can deliver its promise.

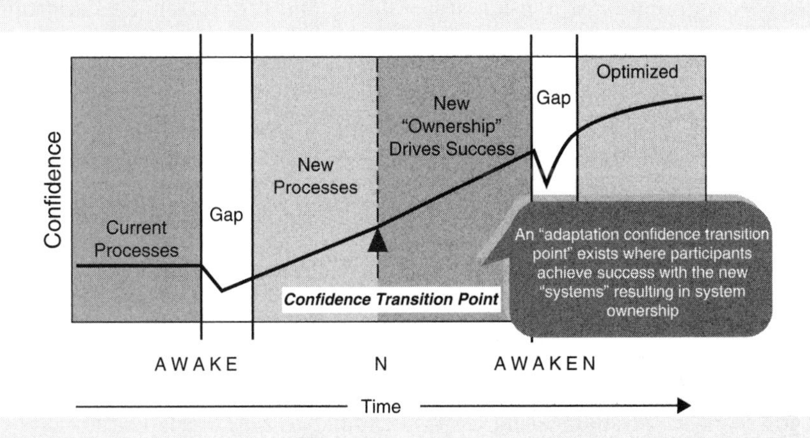

[1] Bloom, B. S., Engelhatt, M. D., Furst, E. J., Hill, W. H., and Krathwohl, D. R., *Taxonomy of Educational Objectives: Handbook I – Cognitive Domain*, McKay, New York, 1956.

[2] Kaplan, R. S., and Norton, D. P., *The Balanced Scorecard: Translating Strategy Into Action*, Harvard Business School, Cambridge, Massachusetts, 1996.

[3] Palmer, I., King, A. W., and Kelleher, D., Listening to Jack: GE's Change Conversations With Shareholders, *Journal of Organization Change Management*, Vol. 17, No. 6, December 2004, 593-614.

– **A. Rhoades, The University of Liverpool**

and to teach the personal and interpersonal skills, and product, process, and system building skills specified in the Syllabus. There must be a process for supporting faculty as they enhance their competence, in teaching, in new forms of evaluation, and in engineering practice and related skills.

Enhancement of faculty competence must be accomplished while protecting the academic careers of faculty. Professional development activities should enhance their opportunities for promotion and tenure, not put future academic promotion at risk. Consistent with the key factors above, the recognition and incentives for faculty ideally should be in support of this approach to professional development.

Enhancement of faculty competence in skills

CDIO programs should provide support for faculty to improve their individual competence in personal and interpersonal skills, and product, process, and system building skills as described in the Syllabus. The nature and scope of faculty development varies with the resources and intentions of each program and institution. Examples of actions that enhance faculty competence include: professional leave to work in industry, partnerships with industry colleagues in research and education projects, inclusion of engineering practice as a criterion for hiring and promotion, and appropriate professional development experiences at the university. Enhancement of faculty competence in skills related to the CDIO Syllabus is the focus of Standard 9.

STANDARD 9 – ENHANCEMENT OF FACULTY SKILLS COMPETENCE
Actions that enhance faculty competence in personal and interpersonal skills, and product, process, and system building skills.

If faculty members are expected to teach a curriculum of personal and interpersonal skills, and product, process, and system building skills integrated with disciplinary knowledge, they need to be competent in those skills themselves. Most engineering professors tend to be experts in the research and knowledge base of their respective disciplines, with only limited experience in the practice of engineering in business and industrial settings. Faculty need to enhance their engineering knowledge and skills so that they can provide relevant examples to students and also serve as role models of contemporary engineers. Faculty development and support has three basic approaches.

- Hire new faculty who have industrial experience or give newly hired faculty a year in industry to gain the experience before they begin teaching.
- Provide educational programs, such as seminars, workshops, and short courses, for current faculty, or allow current faculty leave or sabbaticals to work in industry.
- Recruit senior faculty with significant industry experience to teach and mentor other faculty, or attract practicing engineers from industry to spend time teaching at the university.

Each of these three approaches is described below.

In hiring new faculty, one would consider whether they have had any actual engineering experience. If so, this should be valued as a positive aspect of their background. If not, the department or program could offer released time to fill in this professional experience. As an example, some programs send newly hired faculty to work with industry for one year prior to the start of their formal teaching responsibilities. This program is aimed at professionals beginning their faculty careers immediately after their advanced degrees. The goal of the year with industry is to develop product, process, and system building skills, as well as to broaden their perspectives on engineering research. This time does not count toward the time required to gain promotion. As an added benefit, they return with a deeper understanding of the research needs of industry. Programs must have institutional support to resource this effort.

Programs also face the challenge of encouraging existing faculty to teach personal and interpersonal skills, and product, process, and system building skills in their courses. A variety of approaches can lead to enhanced skills of existing faculty. One approach is to sponsor short courses or training programs within the university on personal and interpersonal skills, and product, process, and system building skills. Commercially available short courses can be used as well. Larger industrial enterprises often have extensive internal training programs and will allow local faculty to participate *pro bono*. Encouraging such programs also sends the message to faculty that program leaders consider these skills important and are willing to expend resources to help faculty acquire them.

Faculty leaves and sabbaticals that are often taken at other universities or in government agencies, can be taken in industry. Again, program leaders must ensure this time is used to expand the faculty member's competence in teaching the CDIO Syllabus skills (Sections 2, 3, and 4). Otherwise, faculty might be inclined to pursue only their research interests.

Finally, programs can attract distinguished engineers with significant experience in product development and system building. Programs will need institutional support for this effort, as career engineers often do not satisfy traditional hiring criteria. An excellent example of a nationally sponsored effort of this type is the *Visiting Professors' Scheme*, sponsored by the Royal Academy of Engineering in the United Kingdom. (See Box 8.4) This program brings experienced engineering professionals back to the university to share their experiences with both students and faculty.

As another example, at MIT a position called *Professor of the Practice* was created to allow the appointment of similar distinguished practitioners. Another approach is to attract senior engineers to short-term placements, such as visitors or adjunct positions. These senior practitioners bring personal and interpersonal skills, and product, process, and system building skills not only to the classroom, but also to their interactions with other program faculty. Consequently, the proficiency level of the entire faculty increases as a result of hiring practiced engineers.

BOX 8.4. EDUCATING ENGINEERS IN DESIGN

The Visiting Professors' Scheme, sponsored by the Royal Academy of Engineering, fosters industry-academia links and aims to help universities teach engineering design to undergraduates in a way that is related to real professional practice. There are three strands to the program:

- Visiting Professors in the Principles of Engineering
- Visiting Professors in Engineering Design for Sustainable Development
- Visiting Professors in Integrated System Design

The Visiting Professors' Scheme was established in 1989, and since then has provided impetus and resources, along with much needed up-to-date industrial experience. There are currently about 120 VPs, working with 46 universities, bringing a wide perspective and practical experience of design to undergraduates and postgraduate courses in engineering education. Experienced engineers, such as the VPs, have much to convey to young engineers, and in passing on their experience there is also the opportunity for the VPs themselves to appreciate more deeply the nature of their work and the processes they use.

In those universities where design education needed to have its role enhanced and the curriculum updated, the VPs have made significant contributions by raising the profile of design, establishing new design-teaching frameworks and courses, and setting up multi-disciplinary case studies and projects in collaboration with industry. The VPs have acted as advisors and reviewers, promoters and negotiators, motivators and inspirers, as well as mentors and assessors. The VP Scheme has endorsed the following three concepts that underpin good practice:

- Design provides an integrating theme for the study of engineering.
- Multi-disciplinary team projects are the best way to introduce students to the technical and organizational complexities of design.
- Much is gained from undertaking these projects in collaboration with industry where appropriate design provides a strong motivation for students to study engineering science with interest and enthusiasm.

– C. PEARCE (ED.), THE ROYAL ACADEMY OF ENGINEERING

Enhancement of faculty competence in teaching and assessment

Programs should support the faculty as they improve their competence in integrated learning experiences, active and experiential learning, and assessment of student learning. Teaching approaches and assessment methods are described in Chapters Six and Seven, respectively. Examples of actions that enhance faculty competence include: support for faculty participation in university and external faculty development programs, forums for sharing ideas and best practices, and emphasis on effective teaching skills at performance reviews and hiring. Enhancement of faculty competence in teaching and assessment is the focus of Standard 10.

STANDARD 10 – ENHANCEMENT OF FACULTY TEACHING COMPETENCE

Actions that enhance faculty competence in providing integrated learning experiences, in using active experiential learning methods, and in assessing student learning.

If faculty members are expected to teach and assess in new ways, they need opportunities to develop and improve their competence in these domains. There are two common approaches to this development task. Many universities have faculty development programs and groups that support improvement in faculty teaching and are often eager to collaborate. In addition, programs seeking to emphasize the importance of teaching, learning, and assessment, should be prepared to commit adequate resources for faculty development in these areas.

Transforming the faculty requires changes, not only to curriculum, but also to teaching and assessment methods. Changing teaching methods is often more threatening to faculty than changing the curriculum. We recognize these fears, and try to reduce or remove barriers to implementing active and experiential learning in the classroom. Bonwell and Sutherland [6] identify five major barriers as

1. lack of coverage
2. increased faculty preparation time
3. large class sizes
4. lack of resources
5. risk to the faculty member

Lack of coverage is the concern that "all of the material won't be covered." This concern is partially overcome by emphasizing student learning rather than faculty teaching. Recent changes in accreditation criteria, focusing on program outcomes rather than on program content, support this effort. Whenever possible, program leaders should provide faculty with compensatory time in order to plan and implement changes to their teaching. Giving faculty time and resources to enhance their teaching competencies accomplishes two objectives: faculty have the necessary time to plan and pilot changes; and it sends the message to the entire faculty that these changes are important and valued. Program leaders, working with senior leaders of the university, need to convince the institution as a whole that faculty efforts to improve teaching and assessment are worthy of inclusion in promotion credentials.

Program leaders can also influence change in the teaching culture during the hiring process. Candidates for faculty positions are usually questioned about their education, research, and job experience, but rarely about their understanding of, and interest in, teaching. Including questions about teaching philosophy, teaching experience, and willingness to experiment with new teaching methods helps to identify candidates who can contribute to implementing CDIO. Prospective faculty can even be asked to give a seminar on education, along with their traditional seminar on research. At Queen's University, in Canada, prospective faculty are asked to teach a mock class to a group of faculty who take on the role of students.

Program leaders can also enlist the support of external education experts through seminars, workshops, and guest lectures. For example, the CDIO

Initiative offers workshops on active and experiential learning to faculty who are interested in enhancing their teaching and assessment skills. Also, most universities have teaching and learning centers, staffed by education experts who are excellent sources of information and support. In some countries, there are national trends toward requiring certain training and educational certification for new university instructors. A good example of a university-wide system to enhance faculty competence in teaching and assessment, found at the Technical University of Denmark, engages every new faculty member in his or her first year at the university. (See Box 8.5)

By using the resources available to them, programs can go about systematically enhancing the competence of the faculty in engineering skills, as well as active and experiential learning and student assessment.

BOX 8.5. ENHANCING FACULTY TEACHING SKILLS AT THE TECHNICAL UNIVERSITY OF DENMARK

Mandatory teaching training for new assistant professors at the Technical University of Denmark (DTU) includes supervision, relevant teaching tasks, a teaching portfolio, a supervisor statement on teaching abilities, and *Education in University Teaching at DTU* (UD_TU).

The program, called UD_TU, is conducted by the small permanent staff at the Learning Lab at DTU in collaboration with a network of associated pedagogical coordinators from the different departments. UD_TU is a practical education focused on student learning. It spans one-and-a-half years with a total workload of 250 hours. UD_TU is based on active learning and action research; the four modules are a mixture of seminars and assignments. The seminars include short introductions, examples, exercises, and group work. The final project concerns the development of participants' own courses in a structured way, that is, trying out new learning activities, getting feedback from students and colleagues, and evaluating the results. Specific activities include:

- Testing the students' background
- Setting up learning objectives
- Planning and implementing teaching
- Assessment that supports learning
- Peer coaching and student evaluation
- Testing the students' understanding

UD_TU has not been designed specifically to meet the CDIO criteria, but in many ways it does, since the focus is on giving students broad engineering competencies. More importantly, the idea behind UD_TU is much like that of CDIO. The participants conceive the idea of a course, design the teaching sequence, implement the sequence by giving the course, and evaluate the experience.

The first UD_TU module, a basic course in learning and teaching, is also taken by Ph.D. students. For more experienced teachers, the Learning Lab at DTU offers two workshops for Ph.D. supervisors and a compulsory workshop series for all supervisors to assistant professors. In addition, there are two modules for pedagogical coordinators.

Since 1999, when UD_TU was initiated, nearly 100 faculty members have completed their UD_TU education, and nearly 100 more are presently engaged in the program. The supervisor modules began in Fall 2005. A new theoretical module is scheduled for 2006.

– H. P. CHRISTENSEN, TECHNICAL UNIVERSITY OF DENMARK

RESOURCES TO SUPPORT PROGRAM CHANGE

The CDIO Initiative is not prescriptive. Program leaders must bring about program change to adapt and implement a CDIO approach. We have created a number of approaches and open-source resources to facilitate adaptation and implementation in diverse university engineering programs. The primary approach is based on an application of the engineering design paradigm and is represented as a roadmap for change. Available resources include practical advice, implementation guidelines, instructional materials, and descriptions of transition activities. The CDIO Initiative also supports collaborative development, and sponsors workshops and conferences to exchange ideas and to develop a mutual understanding of best practice.

Engineering design paradigm for the development of a CDIO approach

An engineering design paradigm has been applied to the development of the CDIO approach. This paradigm also serves as a useful roadmap for adaptation and implementation in existing engineering programs. Education can be viewed as a service, which, like other goods and services, can be engineered using the methods of product and system development and operation. Using the engineering design paradigm to develop a CDIO program has several distinct benefits.

- It appeals to the professionalism of engineering faculty by evoking positive attributes of their profession—addressing needs, solving problems, developing new approaches, applying quality standards. Casting the change process in an engineering design framework enables faculty to feel more comfortable with change.
- It draws on the established competencies of the faculty. Structuring the change process as an engineering task enables them to apply their expertise, for example, defining requirements, building prototypes, and collecting data.
- It ensures that valid learning outcomes are specified and forms the basis for curriculum development, teaching and learning methods, and assessment plans. The engineering design paradigm guides the ways in which the CDIO approach is documented and codified—a comprehensive goal statement, curriculum structuring and mapping techniques, and a quantified model for continuous improvement. The documentation, processes, and examples are the subject of Chapters Three through Nine.

The process of adapting and implementing the CDIO approach is closely aligned with the phases of the product, process, and system lifecycle, as illustrated in Figure 8.1. The CDIO Initiative uses the techniques of product, process, and system development to structure an educational program based

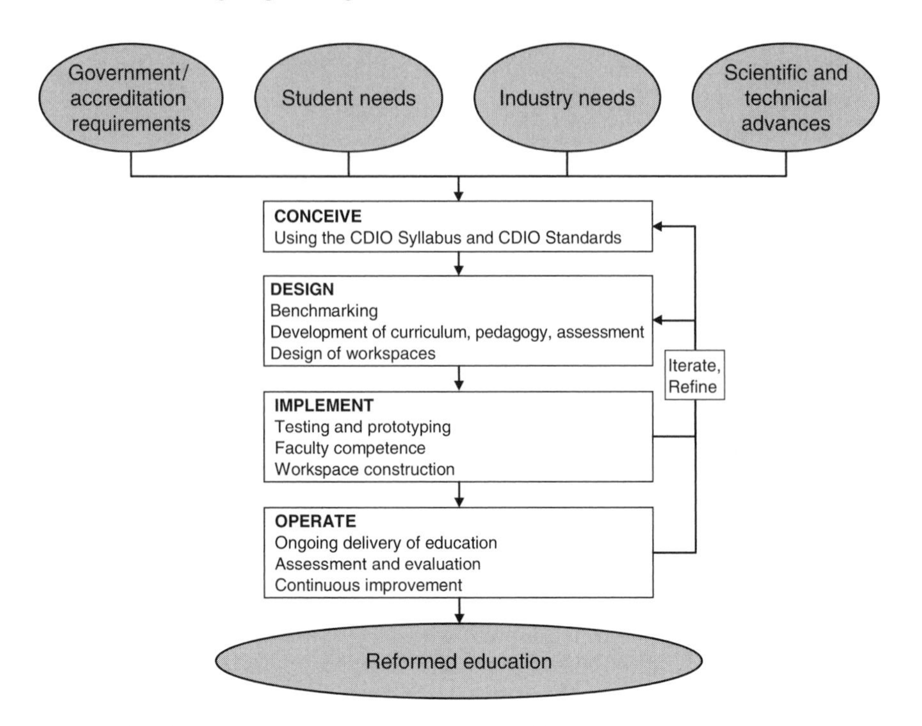

FIGURE 8.1. **DESIGN AND DEVELOPMENT OF A CDIO APPROACH**

on the premise that the proper context for engineering education is the product, process, and system lifecycle. (Figure 2.1 in Chapter Two delineates the generic tasks and outcomes at each phase of the product, process, and system lifecycle.)

The generic *Conceive* phase considers the needs, technology, enterprise strategy, and regulation, then develops the architecture and business case that addresses them. The *Conceive* phase analyzes the needs of a graduating engineer, sets clear and consistent learning outcomes, and works out a concept for engineering education that addresses requirements. This concept for engineering education is consistent with university and national goals and standards, and reflects scientific and technical advances. The outcomes of the *Conceive* phase are unique for each program, but are guided by the CDIO Syllabus, which is the statement of learning outcomes for engineering students, and the CDIO Standards, which are the twelve features that characterize a CDIO program.

The generic *Design* phase involves creating the design—the plans, drawings, and algorithms that describe what will be built or implemented. These activities include benchmarking the existing curriculum, and using open-source tools that aid in curriculum development, course development, teaching and learning methods, assessment methods, and student workspaces. These tools, available on the CDIO website as Implementation Kits (I-Kits) and Instructor Resource Materials (IRM), facilitate adaptation and implementation.

In the generic *Implement* phase, the design is transformed into a product that is tested and validated. New educational tools and resources are tested in engineering programs at collaborating universities. Collaboration enables the CDIO Initiative to compare results, evaluate, iterate, and improve processes and materials, and adapt the approach to engineering programs in a variety of educational environments. It is in this phase that human and physical resources are developed.

Finally, the generic *Operate* phase encompasses the use of the implemented product to deliver the intended value, including maintaining, evolving, and retiring the system. Operation occurs when the educational reform initiatives move beyond prototype and test stages to a steady-state phase where major program changes have been implemented. Evaluation and continuous improvement of the program and of the CDIO approach continue in this phase and are supported by assessment and evaluation.

The process of transitioning to a CDIO program is illustrated in more detail in Figure 8.2. This expansion of the simple view presented in Figure 8.1 broadly retains the major phases, based on the product, process, and system lifecycle. Here, the *Conceive* phase consists of the adoption of Conceiving-Designing-Iimplementing-Operating as the context and the customization of the Syllabus. This is followed by a *Design* phase, comprised of the benchmarking of the program's content and resources, and comparisons with goals and other programs. This phase results in the identification of required changes to curriculum, workspaces, teaching and learning practices, learning assessment, and program evaluation. The *Implement* phase is supported by open-source ideas and resources and presented below. Finally, the *Operate* phase includes the operation and evaluation of the program. The columns in the flowchart align with four themes: curriculum, workspaces, teaching and learning, and student assessment and program evaluation. The numbered boxes in the flowchart represent application of the 12 CDIO Standards.

Open-source ideas and resources

The CDIO Initiative has developed resources to facilitate adaptation and implementation of a CDIO approach. In some cases, these materials are the direct result of our comparative experiences; in some cases they are based on research and scholarship conducted by members of our Initiative; and in still other cases they are adapted from best practice elsewhere. We have tried to assemble a set of resources for the comprehensive reform of a program.

The CDIO approach has been implemented in diverse universities, disciplines and nations. These existing programs also reflect diversity in goals, students, financial resources, existing infrastructure, university context, industry desires, government regulations, and professional societies' accreditation standards. To accommodate this diversity, the approach is codified as an open source. This open, accessible architecture for program materials promotes the dissemination and exchange of ideas and resources.

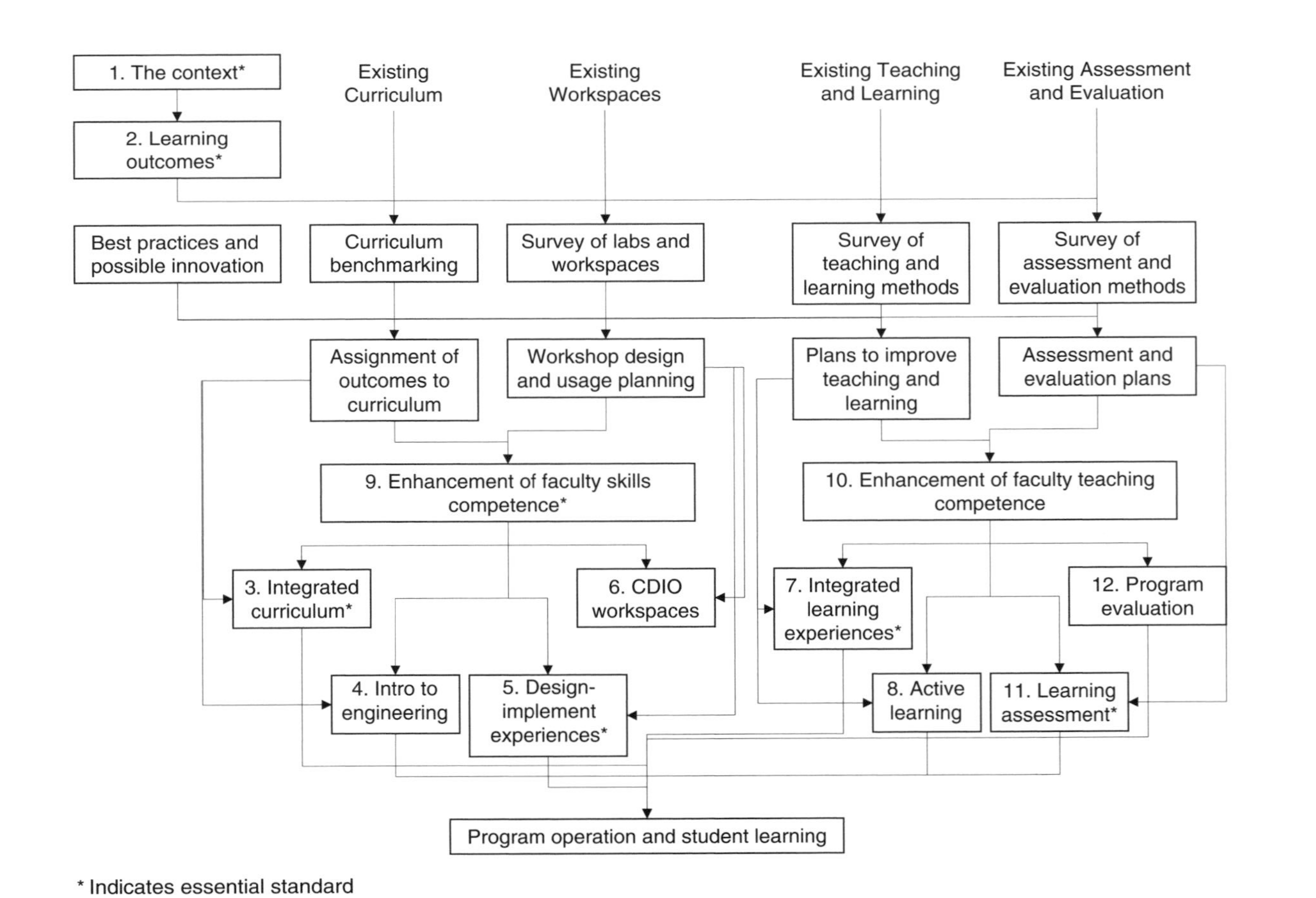

Figure 8.2. The Transition to a CDIO Approach

TABLE 8.2. OPEN-SOURCE RESOURCES OF THE CDIO INITIATIVE

Resource	Purpose	Description
The CDIO Syllabus	Facilitate the creation of clear and comprehensive goals and learning outcomes for engineering programs	A generic, customizable template of goal statements that include technical knowledge and personal and interpersonal skills, and product, process, and system building skills
The CDIO Standards	Distinguish a CDIO program and its graduates, and guide adoption and implementation	Twelve features that characterize a CDIO program, including descriptions, rationale, and evidence that the feature is in place
Start-Up Guidance	Provide ideas and support to program leaders who are adapting and implementing a CDIO approach	Practical advice on how to initiate educational reform and develop a CDIO approach
Implementation Kits (I-Kits)	Provide information, tools, models, and templates to help programs with adaptation and implementation	Guidelines, reports of best practice, tools, and templates in four areas: curriculum, workspaces, teaching and learning, and assessment
Instructor Resource Materials (IRM)	Develop sharable, modifiable teaching resources for faculty	Online, multimedia, instructor guides containing teaching suggestions and assessment tools on specific skills learning outcomes
Published Papers and Reports	Document educational changes accomplished in the CDIO Initiative	Journal and conference papers and reports written by collaborators about the development and adoption of a CDIO approach
CDIO website	Provide information about the CDIO Initiative and its current activities, and archive development activities	Includes tools and resources, CDIO Standards, CDIO Syllabus, papers, information about meetings

A wide variety of resources is available. The major resources are described in Table 8.2. The CDIO Syllabus and CDIO Standards are discussed in detail in Chapters Three and One, respectively. Some other key resources include:

- *Start-up Guidance* is designed for program leaders who are considering or have decided to adapt and implement a CDIO approach. It gives practical advice on how to begin the early steps.
- *Implementation Kits* (I-Kits) are intended for the working groups who will plan and implement reform in the program. These kits are topical, organized around four themes and the Standards, and contain the first information necessary to reform curriculum, plan design-implement experiences, and other information necessary to begin implementation.

- *Instructor Resource Materials (IRM)* are developed for individual faculty who must integrate personal and interpersonal skills, and product, process, and system building skills into their courses or modules. These materials are described in more detail in Chapter Six.
- *Published Papers and Reports* are journal and conference papers and reports written by our collaborators, document development and best practices.
- *CDIO website* is the online repository for the open-source resources, and provides information about current activities of the Initiative.

These and other resources have been developed to enable engineering programs to adapt the CDIO approach to their specific needs. Engineering programs can implement the entire approach or choose specific components.

The CDIO Initiative does not attempt to be prescriptive. We understand that every program, school, university, discipline, and nation has unique needs. We have created a set of resources, approaches, and ideas that can be adapted to improve engineering education for students and other key stakeholders. The materials referenced here, and many of the approaches outlined in other chapters, are still under development. We acknowledge that almost every quality engineering education program is engaged with some, or many, aspects of this reform, and are making contributions from which we can all learn.

Resources are limited at all universities. Very few universities can engage in a comprehensive reform with the expectation that steady-state resources will increase. We have developed the CDIO approach with the assumption that steady-state resources will not increase—that programs will have to retask existing human, space, and time resources. Occasionally, universities and national agencies enable programs to argue for additional steady-state resources. Having a well-developed approach often enables a program to compete more effectively for these new resources.

In the transition from a conventional engineering program to a CDIO program, some additional time and support will be needed. It will not be possible to design a new curriculum, develop new learning experiences, retask workspaces, and develop assessment tools without additional time and effort. We have created the resources outlined here in an effort to minimize this transition, but the effort may still be considerable.

Value of collaboration for parallel development

We have observed that engineering educators worldwide face similar issues. Many of the underlying issues are traceable to the tension between the two main goals of deep understanding of the fundamentals and competence in broader professional skills. Other issues are as common as how one divides students into groups on projects, and assesses these group projects. Addressing even the common issues in a rigorous way, to say nothing of the endeavor to reform the entire education, is a challenge for any single program or department. The scope of these issues, and their worldwide commonality, suggest that it would be desirable to work together to address them, in an organized way toward common goals. The international CDIO Initiative enables us to develop and implement a general and adaptable educational model. The value of international collaboration lies in

- Creating more robust and generalized starting positions for the development of CDIO programs. For example, surveys and benchmark studies compare stakeholder expectations and institutional conditions at different universities and countries
- Sharing approaches and ideas within a structured framework of common goals
- Creating a set of transferable resources, including Implementation Kits (I-Kits), Start-Up Guidance, Instructor Resource Materials (IRM), that can be used by other universities to facilitate adaptation and implementation
- Sharpening of key features of the CDIO Standards

There is another important benefit of working in a collaborative format. Working together promotes visibility into what others are doing. This raises the level of ambition for educational development at our universities through friendly competition toward a mutual goal. It strengthens the arguments for adopting a CDIO approach at collaborators' respective universities. Evidence of success at competitors' universities can be a very persuasive force for change.

In order to facilitate interaction, the Initiative sponsors a number of activities. These include:

- *Workshops* at which the key ideas are introduced
- *Regional Meetings* that allow participating universities within a geographic region to come together to exchange ideas and develop new approaches
- *Annual International Meetings* that bring together educators from around the world to exchange key learning and successes that have emerged

The Initiative invites participation in all these activities, as well as contributions from programs engaged in the improvement of engineering education.

SUMMARY

The transformation to a CDIO program touches all faculty members in a department or program, as it resets the context and reorganizes the education process. This chapter examined twelve key success factors that facilitate organizational change, and gave examples of how engineering programs incorporated these factors in their transition. The importance of faculty support and development was also highlighted. It showed how programs wishing to implement a CDIO approach to engineering education can take advantage of the experiences of existing programs, and use the open-source resources available on the website. The next chapter creates a program evaluation framework based on the twelve CDIO Standards.

DISCUSSION QUESTIONS

1. What strategies have you employed to implement change? Are they aligned with those suggested in this chapter?
2. What policies and incentives are in place to enhance faculty competence? How effective have they been?
3. Can you identify people, programs, and resources that support faculty development in teaching and assessment at your university?
4. How do you anticipate using the resources provided by the CDIO Initiative?

References

[1] Rudolph, F., *The American College and University: A History*, The University of Georgia Press, Athens, Georgia, 1990.

[2] Bush, V., *Science The Endless Frontier*, U. S. Government Printing Office, Washington, D. C., 1945.

[3] *The Bologna Accord.* Available at http://en.wikipedia.org/wiki/Bologna-process

[4] Gleicher, D., Beckhard, R., and Harris, R., *Change Model Formula*, 1987. Available at http://www.valuebasedmanagement.net

[5] Kotter, J. P., Leading Change: Why Transformation Efforts Fail, *Harvard Business Review*, March-April, 1995.

[6] Bonwell, C. C., and Sutherland, T. E., "The Active Learning Continuum: Choosing Activities to Engage Students in the Classroom", in Sutherland, T. E., and Bonwell, C. C. (Eds.), *Using Active Learning in College Classes: A Range of options for Faculty*, New Directions of Teaching and Learning, No. 67, Jossey-Bass, San Francisco, California, 1996.

CHAPTER NINE
PROGRAM EVALUATION

WITH P. J. GRAY

INTRODUCTION

In previous chapters, we described key characteristics of a CDIO program. First, we addressed *what* we should teach: learning outcomes that address disciplinary content, as well as personal and interpersonal skills, and process, product, and system building skills. We went on to discuss *how* we should teach: an integrated curriculum; a sequence of design-implement experiences in workspaces specifically designed to support them; integrated teaching, learning, and student assessment. In Chapter Eight, we presented approaches to enhance faculty competence in these skills and in teaching methods. We now address three key questions dealing with the effectiveness of our approach:

- *How do we determine if programs are successfully implementing a CDIO approach?*
- *How can we improve programs that are not up to standard?*
- *What is the impact of implementing a CDIO program?*

These are the general questions of program evaluation. We define program evaluation as a process for judging the overall effectiveness of a program based on evidence of progress toward attaining its goals. The specific approach to program evaluation can take a variety of forms, depending on the conceptual framework and rationale for the evaluation. Evaluation of CDIO programs follows primarily a judgment model, based on inputs, processes, and outputs. *Inputs* include feedback from personnel, use and usability of facilities, and use and availability of resources; *processes* include teaching, assessment, evaluation methods; and, *outputs* are the intended learning outcomes for students and overall program outcomes.

We evaluate a program by judging its overall quality based on indications of its progress toward reaching its goals. One way to judge this overall quality is to focus on a program's progress toward implementation of the 12 CDIO Standards described throughout this book. Because the Standards address inputs, processes, outcomes, and to a limited extent, impact, program evaluation based on the CDIO Standards can provide program leaders with data upon which to determine whether programs are achieving their goals, operating effectively, allocating resources appropriately, and making a difference overall.

We use the term *standards-based program evaluation* to describe the approach we use with CDIO programs. This approach is consistent with a judgment model of program evaluation. A standard, in this context, is a criterion or characteristic that defines a program. Evidence of progress toward implementation of a CDIO approach is collected from multiple sources, using a variety of quantitative and qualitative methods. When this evidence is regularly reported back to faculty, students, program administrators, alumni, and other key stakeholders, the feedback forms the basis for making decisions about the program and its continuous improvement.

Standards-based program evaluation, using the 12 CDIO Standards, is consistent with accreditation models and other national evaluation approaches. This consistency is based on similar purposes. Both approaches set criteria, collect evidence of compliance with the criteria, and require plans to improve programs. However, national accreditation criteria often establish minimum levels of acceptable program performance. We have deliberately set the CDIO Standards high, so that all programs—even those of the highest quality—can use them as the basis of continuous improvement.

In this chapter, we discuss the purpose and value of a standards-based approach to program evaluation as a way to determine if programs are successfully implementing a CDIO approach. In doing so, we identify key evaluation questions aligned with the Standards, and examine a variety of methods to collect data to answer the evaluation questions. We give examples of data collection and analysis in representative programs. We make connections of program evaluation results with the process of continuous improvement and give suggestions for improving programs that are not up to standard. Finally, we summarize results that give evidence of the impact of CDIO programs overall.

CHAPTER OBJECTIVES

This chapter is designed so that you can

- recognize the characteristics of a standards-based approach to program evaluation
- identify key questions that guide program evaluation and align them with the CDIO Standards
- describe a variety of methods that provide evidence of program quality
- give examples of standards-based program evaluation
- emphasize the connection between program evaluation and continuous program improvement
- evaluate the overall impact of programs that have implemented a CDIO approach

STANDARDS-BASED PROGRAM EVALUATION

The conceptual framework of program evaluation depends on the purpose and rationale for conducting the evaluation. For example, objectives-based models focus on the purpose of the program and the attainment of specified goals, objectives, and outcomes. In contrast, goal-free evaluation focuses on the outcomes without the specification of any pre-determined goals. Naturalistic approaches are broad in scope, focusing on human elements and processes of the program in specific contexts. Judgment models, such as program accreditation, address compliance with standard guidelines, and tend to focus on inputs and processes. Management-oriented program evaluation focuses on key questions of decision makers, limiting the range of data collection to specific questions. These questions tend to emphasize the outcomes and overall impact of a program [1].

Evaluation of CDIO programs is primarily a judgment model, with components of objectives-based and management-oriented models. Similar to many accreditation models, judgments are made on the inputs to the program, for example, qualifications of the academic staff, access to modern engineering tools, workspaces, and program processes, for example, teaching, advising, enrollment. Program evaluation, then, is a matter of showing compliance with criteria that address these inputs and processes. In recent years, accreditation models have broadened their scope to include outcome measures. Similar to objectives-based models, program evaluation focuses on the attainment of program goals and specific program learning outcomes. Management-oriented models, such as *The Balanced Scorecard* [2] implemented at Linköping University, contribute components of strategic planning, allocation of resources, and measurement of impact, all of which broaden program evaluation beyond judgment and objective-based models.

Standards-based program evaluation describes any approach to evaluation that focuses on the explicit criteria, standards, and other components of the evaluation process [3]. This approach aligns well with the rationale for the three models presented—judgment, objectives-based, and management-oriented—and shares common features with them. Standards that address program objectives and outcomes focus on the end results of the program for the people it is intended to serve. These standards concern the cumulative results of the educational experiences offered to students by the program. Included are student learning outcomes in courses and other activities, as well as the culminating outcomes that are expected as a result of completing a program. Of course, it is hoped that most of these outcomes are the ones intended, but there also may be unintended outcomes.

In addition to outcomes, standards-based program evaluation examines the processes that lead to those outcomes. Process evaluation is the systematic review of what is happening inside the program and involves an evaluation of how the program is operating in order to meet its goals. Program processes may include admissions, advising, registration, student support services, teaching, learning, and internship and job placement. Examination of these processes helps to explain the program outcomes and points to features of the program that are more or less successful [4].

To a limited extent, standards-based program evaluation also measures the overall impact of a program, by looking at what happens to participants and others as a result of the program. Such impact is often construed as long-term outcomes and may include the effects of the program on the larger community and society. Impact studies may look at workforce capabilities, ethnic and gender equality, and productivity. Such studies may follow graduates for their entire careers to determine the long-term impact of a program [4].

In evaluating a program within the framework of the CDIO Standards, we examine evidence of processes and outcomes, and to a limited extent, inputs and impact. Taking a broad view, Standards 1 and 6 address inputs; Standard 2 specifies the intended learning outcomes; Standards 3, 4, 5, 7, 8, 9, 10, and11 focus on processes. While the Standards do not specifically address long-term impact, the evaluation of our programs often includes questions related to students' future plans, alumni contributions to their engineering fields, and influences of a program on local, national, and international industries. The remaining standard, Standard 12, is the criterion for program evaluation itself, that is, a CDIO program takes a systematic and comprehensive approach to data collection and analysis and program improvement.

STANDARD 12 – PROGRAM EVALUATION

A system that evaluates programs against these twelve standards, and provides feedback to students, faculty, and other stakeholders for the purposes of continuous improvement.

Standards-based evaluation is systematic in that it identifies and addresses a wide range of questions, uses a variety of methods to collect and analyze data, and uses the data to make decisions about program effectiveness and the need for continuous improvement. We now examine this systematic evaluation process as it applies to CDIO programs.

THE CDIO STANDARDS AND ASSOCIATED KEY QUESTIONS

Program evaluation is based on the 12 Standards that we have developed. Before proceeding with a detailed discussion of the process of program evaluation, we present the rationale and organization of the standards themselves. The Standards were introduced in Chapter Two, and discussed as the organizing theme of Chapters Three through Eight. They are listed in Appendix B, with a description and rationale for each standard, and examples of evidence that the standard is in place.

Rationale and organization of the CDIO Standards

The CDIO Initiative developed and adopted the Standards in order to help programs as they address the perceived need—to educate students who are able to conceive, design, implement and operate complex value-added engineering products, processes, and systems in a modern, team-based environment. The Standards form a bridge from the program goals to a tangible set of educational inputs, processes, and outcomes. They give guidance to individual university programs regarding how to proceed, and attempt to answer the central questions of engineering education reform:

- *What is the full set of knowledge, skills, and attitudes that engineering students should possess as they leave the university, and at what level of proficiency?*
- *How can we do better at ensuring that students learn these skills?*

It is important to understand what the Standards are, and are not. They are a means of guiding programs toward fulfillment of specific needs and goals. They are a codification of best practice based on research and our collective experience around the world. They are intended to provide support for change in the direction desired by program stakeholders. They are designed to be consistent with national accreditation and evaluation criteria, yet also provide international benchmarks against peer institutions. In addition, the Standards form the basis for program evaluation and continuous improvement.

The Standards are intended to distinguish those programs which offer a comprehensive CDIO approach to education from those who incorporate only a few of the components. They were originally developed in response to program leaders, alumni, and industrial stakeholders who wanted to know how they would recognize CDIO programs and their graduates. Of the 12 Standards, we distinguish seven that are essential to the CDIO approach:

- Standard 1 The Context
- Standard 2 Learning Outcomes
- Standard 3 Integrated Curriculum
- Standard 5 Design-Implement Experiences
- Standard 7 Integrated Learning Experiences
- Standard 9 Enhancement of Faculty Competence
- Standard 11 Learning Assessment

The remaining five Standards are considered supplementary and indicative of good practice, but not necessarily distinguishing features of a CDIO program: Standard 4 Introduction to Engineering; Standard 6 Engineering Workspaces; Standard 8 Active Learning; Standard 10 Enhancement of Faculty Teaching Competence; and Standard 12 Program Evaluation.

The Standards do not include all components of an engineering program. They omit some of common inputs and processes, for example, faculty qualifications in their engineering disciplines, student advising and counseling, and classroom facilities. This limited scope is deliberate in order to accentuate

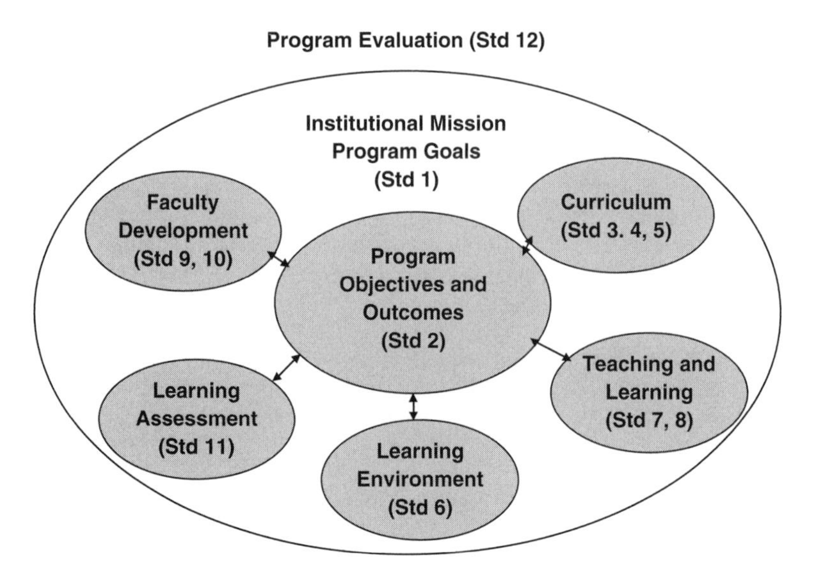

FIGURE 9.1. **PROGRAM EVALUATION AND THE CDIO STANDARDS**

distinctions and bias toward a program vision that has been found to meet the program goals. Having said this, there is nothing in the Standards that is absolutely unique to programs that are members of the CDIO Initiative. High-quality engineering education programs around the world have some of the inputs, processes, and outcomes discussed in the Standards, and programs of the highest quality have many. The Standards are offered as a guide for continuous improvement, regardless of a program's initial quality.

The Standards can be organized in a way that is consistent with traditional program evaluation. As illustrated in Figure 9.1, program evaluation focuses on the outcomes of the program and the processes that contribute to students' achieving those outcomes, as embodied in the 12 Standards. The Standards can be grouped into one or more focus areas: program mission and goals, curriculum, teaching and learning methods, the learning environment, learning assessment, and faculty development. Note that program evaluation, itself, is one of the standards.

Key questions aligned with the Standards

In planning an evaluation, key questions are posed for each important focus area. Table 9.1 illustrates the alignment of the key questions of the evaluation plan with the 12 CDIO Standards. These key questions are derived from the descriptions of the CDIO Standards.

TABLE 9.1. **KEY QUESTIONS ALIGNED WITH THE CDIO STANDARDS**

Key questions
Institutional Mission and Program Goals
Standard 1 – The Context
• To what extent do the institutional mission and program goals reflect the adoption of the principle or program philosophy that product, process, and system lifecycle development and deployment—Conceiving, Designing, Implementing, and Operating—are the context for engineering education?
• To what extent is the product, process, and system lifecycle considered the context for engineering education in that it is the cultural framework, or environment, in which technical knowledge and other skills are taught, practiced and learned?
Program Outcomes
Standard 2 – Learning Outcomes
• To what extent are specific, detailed learning outcomes for personal and interpersonal skills, and product, process, and system building skills, consistent with program goals and validated by program stakeholders?
• How have stakeholders helped to determine the expected level of proficiency, or standard of achievement, for each learning outcome?
Curriculum
Standard 3 – Integrated Curriculum
• How are learning outcomes for personal and interpersonal skills, and product, process, and system building skills integrated into the curriculum?
• To what extent is the curriculum designed with mutually supporting disciplinary courses, with an explicit plan to integrate personal and interpersonal skills, and product, process, and system building skills?
Standard 4 – Introduction to Engineering
• How effectively does the introductory engineering course provide the framework for engineering practice in product, process, and system building skills, and introduce essential personal and interpersonal skills?
• To what extent do introductory courses stimulate students' interest in, and strengthen their motivation for, the field of engineering by focusing on the application of relevant core engineering disciplines?
Standard 5 – Design-Implement Experiences
• Does the curriculum include two or more design-implement experiences, including one at a basic level and one at an advanced level?
• How are opportunities to conceive, design, implement, and operate products, processes, and systems included in the required curriculum and elective co-curricular activities?
Teaching and Learning
Standard 7 – Integrated Learning Experiences
• Are there integrated learning experiences that lead to the acquisition of disciplinary knowledge, as well as personal and interpersonal skills, and product, process, and system building skills?
• How do integrated learning experiences incorporate professional engineering issues in contexts where they coexist with disciplinary issues?
Standard 8 – Active Learning
• How do active and experiential methods contribute to the attainment of program outcomes in a CDIO context?
• To what extent are teaching and learning methods based on approaches that engage students directly in thinking and problem solving activities?

(*Continued*)

TABLE 9.1. KEY QUESTIONS ALIGNED WITH THE CDIO STANDARDS—CONT'D

Key questions
Learning Environment
Standard 6 – Engineering Workspaces
• How do workspaces and other learning environments support hands-on and experiential learning?
• To what extent do students have access to modern engineering software and laboratories that provide them with opportunities to develop the knowledge, skills, and attitudes that support product, process, and system building skills?
• Are workspaces student-centered, user-friendly, accessible, and interactive?
Learning Assessment
Standard 11 – Learning Assessment
• How is assessment of student learning in personal and interpersonal skills; product, process, and system building skills; and disciplinary knowledge embedded in the program?
• How are these learning outcomes measured and documented?
• What have students achieved with respect to program outcomes?
Faculty Development
Standard 9 – Enhancement of Faculty Skills Competence
• How are actions that enhance faculty competence in personal and interpersonal skills, and product, process, and system building skills supported and encouraged?
Standard 10 – Enhancement of Faculty Teaching Competence
• What actions have been taken to enhance faculty competence in providing integrated learning experiences, using active experiential learning methods, and assessing student learning?
Program Evaluation
Standard 12 – Program Evaluation
• Is there a systematic process in place to evaluate programs against the 12 CDIO Standards?
• To what extent are evaluation results provided to students, faculty, and other stakeholders for the purposes of continuous improvement?
• What is the overall impact of the program?

In collecting data to answer these key evaluation questions, it is important to bear in mind four factors related to evidence:

- the criteria of success for each important area, that is, what does a good example look like?
- the evidence that will indicate that the program is doing well in each key area
- the kind of evidence that will persuade key stakeholders
- the way in which the evidence should be summarized for different stakeholder groups

In summary, in a CDIO program, the criteria of success are the 12 Standards. A program is considered successful if it can show evidence that the program components described in the Standards are in place. Different stakeholder groups will emphasize subsets of the 12 Standards, but all Standards are important measures for at least one stakeholder group. In a later section of this chapter, we give examples of ways to document the extent to which each of the key evaluation questions has been answered in representative CDIO programs. We now examine a variety of methods to collect reliable and valid evaluation data.

METHODS TO EVALUATE PROGRAMS

Once the key questions have been articulated, it is important to consider the source of the information and how the data may be collected within resource constraints of the program. Effective program evaluation makes use of multiple data collection methods at multiple points in the program. Some of the methods used in existing programs to determine quality and design plans for continuous improvement are described here. These include document reviews, interviews, surveys, reflective memos, expert reviews, and longitudinal studies. In addition, the assessment section of the CDIO website has examples of tools in many of these categories.

Document reviews

In implementing the Standards, it is helpful to document plans and actions at each step in the process. For example, program mission statements and learning outcomes, curriculum designs, and course syllabi can be archived in order to document program development. Including the analysis of current facilities, teaching-learning methods, and assessment techniques in the documentation helps to identify best practices and areas of potential innovation. Data collected as the program is being implemented can guide its refinement and continuous improvement. Reports of student learning outcome assessment and evaluations of specific program components can provide the data for judging the success of the program in achieving its goals. A document review process focuses on the importance of establishing and maintaining a program archive. In our programs, these documents are reviewed internally, but are not usually shared externally.

Personal and focus group interviews

Formal documents do not tell the whole story of a program's success. Personal and focus group interviews can provide information about the effect of programs on students and other stakeholders. They have the advantage of being able to ask open-ended exploratory questions. We use personal interviews to gather data from students as they begin their programs, and again, as they complete them. Interviews have also been conducted to gather information from faculty members about the specific teaching and assessment methods they use in their courses. Focus group interviews provide more complete data than personal interviews because the group interaction generates additional questions and responses. Some CDIO programs conduct focus group interviews to evaluate courses. Panels of students, faculty, and course managers meet to review each course at the end of each term. Focus groups are also used to gather input from key stakeholders to define program outcomes. Examples of these focus groups were described in Chapter Three.

Questionnaires and surveys

As with interviews, questionnaires and surveys ask a common set of questions. However, they can be more efficient than interviews since they may be used to collect large amounts of information from diverse groups of respondents in the same time period. Moreover, it is possible to collect statistically valid data from large samples. Responses can be collected in person, by postal mail, by email, or online. Examples of questionnaires and surveys in our programs include stakeholder surveys of the CDIO Syllabus, student ratings of faculty and courses, and exit surveys of graduating students.

Instructor reflective memos

In reflective memos, instructors summarize their experiences with teaching, learning, and assessment in their respective courses. Memos may address the following:

- the intended learning outcomes and evidence that they have been met
- the ways in which personal and interpersonal skills, and product, process, and system building skills have been integrated into their courses
- evidence that their teaching and assessment methods have been effective
- their plans to improve the course in subsequent offerings, and
- the names of instructors with whom they will share the memo

These individual memos can be summarized across a program to form another source of information about the success of a program. Instructors can also meet with the program head, or the person responsible for instructional quality, to discuss the memo and other issues related to curriculum and instruction.

Faculty members at MIT have been writing reflective memos since 1999. They report that the underlying value of these memos is the opportunity to reflect on their teaching and record proposed changes while the experience of the term is still fresh in their memories. This practice has made a significant contribution to the improvement of teaching and learning in the program. Annual summaries of these reflective memos are also valuable sources of information for regional and national accreditation reviews.

Program reviews by external experts

It is often helpful to have people who are not directly connected to a program provide an independent evaluation. External evaluators may be provided with questions such as those listed in Table 9.1. In preparation for visiting committees, program personnel prepare summary materials that include

- a program evaluation plan
- information about the program
- a self-study identifying program strengths, weaknesses, and issues related to the Standards
- specific questions that program stakeholders would like to address

Program reviews by external experts include regional and national accreditations, institutional reviews, and certification or approval ratings from engineering professional associations. For example, in Sweden, programs are reviewed by the Swedish National Agency for Higher Education. A description of a recent review is found in Box 9.2 in a later section of this chapter.

Longitudinal studies

In longitudinal studies, data are collected from groups of respondents over time. Data may be collected from cohort groups, that is, students who go through the program together, at regular intervals during and after their involvement in the program, or from different groups who are studied at the same point in their programs. Interviews and questionnaires are common methods for collecting longitudinal data. Linköping University provides a good example of a longitudinal study that examines students' expectations and satisfaction from the time they enter the Applied Physics and Electrical Engineering program until the time they complete their degree requirements. This longitudinal study is described in Box 9.1 in a later section of this chapter.

EVALUATING A PROGRAM AGAINST THE CDIO STANDARDS

Once reliable and valid data have been collected and analyzed, the results are organized in such a way as to answer the key evaluation questions, posed earlier in this chapter. As explained earlier, program evaluation focuses on the 12 Standards. Similar to most judgment models of evaluation, determination of a program's progress toward meeting the Standards may be accomplished through self-evaluation. We use a five-level rating scale to indicate progress toward the planning, implementation, and adoption of each Standard. As seen in the rating levels in Table 9.2, and in the graph in Figure 9.2, planning, implementation, and adoption of the Standards is not a linear process. The rubric has been designed to encourage planning and allow various styles of implementation and adoption. Programs use this rubric for self-evaluation against the 12 Standards at least annually.

TABLE 9.2. **RATING SCALE FOR SELF-EVALUATION**

Rating Scale	Description
0	No initial program-level plan or pilot implementation
1	Initial program-level plan and pilot implementation at the course or program level
2	Well-developed program-level plan and prototype implementation at course and program levels
3	Complete and adopted program-level plan and implementation of the plan at course and program levels underway
4	Complete and adopted program-level plan and comprehensive implementation at course and program levels, with continuous improvement processes in place

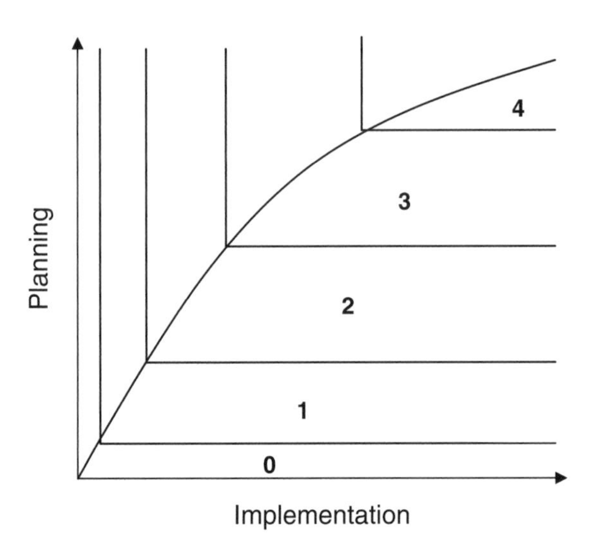

FIGURE 9.2. INTERPRETATION OF RATING SCALE FOR SELF-EVALUATION

In addition to the numerical ratings, each program describes the evidence that is the basis for the rating of each standard. This evidence serves as the foundation for decisions about program improvement. Initially, the programs that were involved in the development of the CDIO approach declined comparisons of ratings across institutions, because of the inherent subjectivity of the self-rating process. The real value of the program self-ratings is their contribution to the continuous improvement process. However, recently the program self-ratings and descriptions of evidence were examined for twelve programs to help determine the overall impact of the CDIO Initiative. Table 9.3 lists a sample of the self-ratings and evidence across 12 programs as of May 2005. The table can be used as a guide for a program to determine its rating for each of the Standards. The examples in the table come directly from the self-reports of the representative programs [5]. In the future, a more formal rubric, based on these examples, will be developed to facilitate self-evaluation.

The factor that accounts for most of the wide variation in the self-ratings is the number of years that a program has been in the CDIO Initiative, or from the start of a program's engineering education reform. The twelve programs included in Table 9.3 range from the four founding institutions to three programs that began their reform efforts less than two years prior to their self-evaluations. A few trends can be seen in the self-reports:

- High-level program and curriculum design is the focus at the start of a program's reform initiatives.
- New methods of teaching, learning, and assessment are implemented after the second or third year of program reform.
- Although CDIO programs begin program evaluation at the start, they need a few years to implement comprehensive systematic program evaluation.

TABLE 9.3. SELF-RATINGS AND EVIDENCE ACROSS TWELVE CDIO PROGRAMS

Rating	Representative examples of evidence
	Standard 1 – The Context
0	(no programs with a rating of 0)
1	The CDIO approach has been accepted as part of the departmental educational development plan
2	A multidisciplinary program that stresses product lifecycle development and deployment is complete, and is scheduled to start next year
3	A new curriculum with a complete program plan based on the CDIO approach was approved by academic council, with implementation started in the first year
4	The process, product, and system lifecycle was adopted as the context by program management, with documentation in the program goal document
	Standard 2 – Learning Outcomes
0	(no programs with a rating of 0)
1	Stakeholders have been consulted, but not yet surveyed with respect to the CDIO Syllabus
2	Detailed learning outcomes exist for the first three years of the program; course learning outcomes have to be reviewed for advanced levels
3	Stakeholder surveys have been completed, and the development of learning outcomes is in progress
4	The CDIO Syllabus has been validated with program stakeholders
	Standard 3 – Integrated Curriculum
0	(no programs with a rating of 0)
1	A multidisciplinary curriculum is being planned, with competencies that include personal and interpersonal skills
2	A framework for integrating skills is found in the project courses; with other parts of the curriculum under development
3	The program has a documented plan that integrates skills with technical disciplinary content
4	An integrated curriculum has been fully implemented; every course has a plan for integrating specific skills
	Standard 4 – Introduction to Engineering
0	(no programs with a rating of 0)
1	A problem-based learning project stresses product and system building, and introduces personal skills
2	Elements of a CDIO approach have been embedded in the 1st-yr design course, with a new year-long introductory course to start next year
3	Two courses and one project in the 1st year serve as the introduction
4	An introductory course is being implemented in the 1st semester, and is documented in the I-Kit on the CDIO website
	Standard 5 – Design-Implement Experiences
0	(no programs with a rating of 0)
1	(no programs with a rating of 1)
2	A sequence of design-implement courses and a number of electives are included in the curriculum plan
3	Four design projects – one each year – are included in the new program; the 2nd-yr and 4th-yr projects are specifically design-implement
4	The program includes two design-implement experiences in addition to the introductory course; 11 courses are currently offered in the 4th year

(*Continued*)

TABLE 9.3. SELF-RATINGS AND EVIDENCE ACROSS TWELVE CDIO PROGRAMS—CONT'D

Rating	Representative examples of evidence
	Standard 6 – Engineering Workspace
0	Inadequate space; lab safety is a concern; engineering tools are out-of-date; workspaces are not student-centered nor user friendly
1	Workspaces exist in limited quantity; new spaces are being planned for the coming year
2	Some spaces have been adapted to C-D-I-O, with further work in progress
3	Most of the in-depth specializations have workspaces and laboratories to support design-implement experiences
4	The learning laboratory has spaces designated for C-D-I-O; workspaces contribute significantly to students' satisfaction with the program
	Standard 7 – Integrated Learning Experiences
0	(no programs with a rating of 0)
1	(no programs with a rating of 1)
2	Integrated learning is prevalent in the 1st-yr program
3	Projects are designed as integrated learning experiences; projects are also integrated with disciplinary courses
4	Problems from industry are used as design assignments and capstone projects; industrial partners are involved in learning experiences
	Standard 8 – Active Learning
0	(no programs with a rating of 0)
1	(no programs with a rating of 1)
2	Some courses use active learning, but more focus needs to be put on this
3	Methods include laboratory work, design activities, experimental learning projects and self-assessment exercises
4	Muddy cards, concept questions, personal response systems, and turn-to-your-partner methods are used in lecture-based courses; problem solving, projects, and experimentation are used in project-based courses
	Standard 9 – Enhancement of Faculty Skills Competence
0	(no programs with a rating of 0)
1	New training programs will be launched by the university's educational and staff development department
2	Current program includes three days of workshops on the new program with emphasis on personal and interpersonal skills
3	Actions include hiring faculty with CDIO expertise, sponsoring faculty to work in industry, and sabbaticals in engineering practice
4	(no programs with a rating of 4)
	Standard 10 – Enhancement of Faculty Teaching Competence
0	(no programs with a rating of 0)
1	Programs being planned by the university education and staff development center
2	Courses offered through university teaching center; mandatory attendance for new faculty
3	Resources available for teaching improvement from the university teaching lab; teaching skills considered during performance reviews
4	Faculty development program in place for more than 25 years; university education experts attend meetings of the CDIO Initiative

TABLE 9.3. SELF-RATINGS AND EVIDENCE ACROSS TWELVE CDIO PROGRAMS—CONT'D

Rating	Representative examples of evidence
	Standard 11 – Learning Assessment
0	(no programs with a rating of 0)
1	Assessment in the final-year project has begun, but a more comprehensive plan is needed
2	Assessment is fragmented within courses; methods include oral exams, peer assessment of projects and presentations, and reflective portfolios
3	Assessment in project courses separates course and project objectives; assessment to a large extent is carried out using the project model
4	Assessment methods are matched appropriately to learning outcomes and reviewed by external examiners and division heads; methods include oral exams, presentations, peer assessment and reflective portfolios
	Standard 12 – Program Evaluation
0	Program evaluation is sporadic and incomplete
1	Program evaluation is at the planning stage
2	An internal system of quality control is used, with plans to measure the program against the CDIO standards
3	Methods include entrance survey, compulsory evaluations, exit surveys, and a national higher education survey
4	Comprehensive system that includes course evaluations, longitudinal studies of student expectations and satisfaction, entry surveys, and alumni surveys

Summarizing program self-evaluations across all members of the CDIO Initiative sheds light on areas in which the collaboration can support member programs.

CONTINUOUS PROGRAM IMPROVEMENT PROCESS

In addition to providing information about a program's progress and status, self-evaluation against the CDIO Standards enables a program to plan specific actions for continuous improvement. Standards relating to input, processes, and outcomes are all examined for areas that fall short of full implementation or that lack the quality desired by each respective program. Figure 9.3 illustrates a continuous program improvement process at four stages:

- Input
- Processes
- Outcomes
- Improvement

At the *Input* stage, a program examines data related to its mission, goals, and purpose, the adequacy of its resources, and the qualifications of faculty and staff. At the *Processes* stage, a program looks at the effectiveness and efficiency of its processes, including teaching, advising, student assessment,

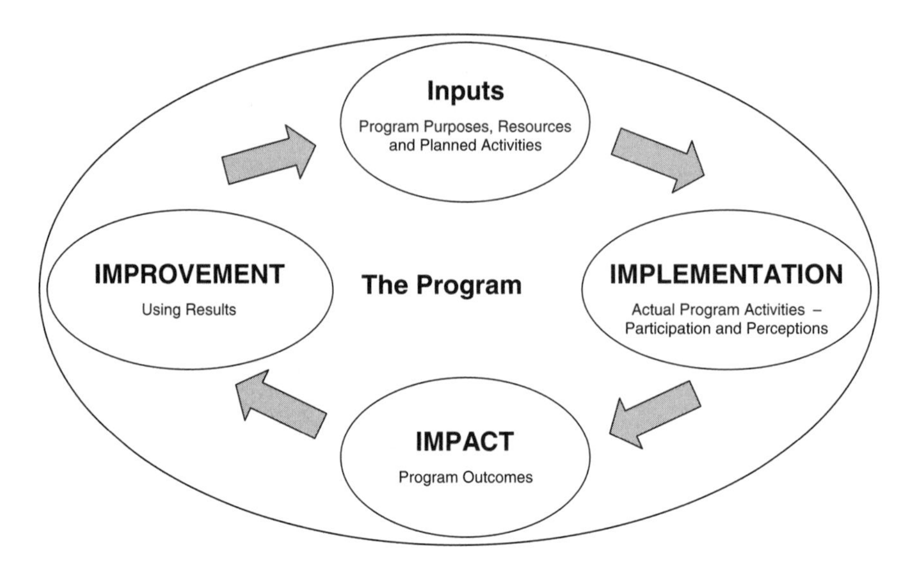

FIGURE 9.3. **CONTINUOUS PROGRAM IMPROVEMENT PROCESS**

and other activities. The *Outcomes* stage focuses on the analysis of results, including short-term learning outcomes of students, as well as long-term impact of the program on stakeholders, the local community, and professional disciplines. The process is not complete, however, until evaluation results from the first three stages are used to improve the program. The cycle is continually repeated, collecting and analyzing data, and using results for program improvement.

To illustrate the way that program evaluation data is used as the basis for continuous improvement and decision making, Box 9.1 describes a longitudinal study at Linköping University [6]. The purpose of the study is to provide reliable and valid data to the program in *Applied Physics and Electrical Engineering* to support its goal of making its program more attractive to students, and particularly to students from under represented populations.

OVERALL IMPACT OF CDIO PROGRAMS

As stated in Chapters One and Two, the overall goals of the CDIO Initiative are explicitly to improve the education of engineering students, and implicitly to educate more engineers. The explicit statement of the goals is—*To educate students who are able to*:

- *Master a deeper working knowledge of technical fundamentals*
- *Lead in the creation and operation of new products, processes, and systems*
- *Understand the importance and strategic impact of research and technological development on society*

Box 9.1. Longitudinal Studies for Program Evaluation at Linköping University

In the 1997–1998 academic year, the program board of Linköping University's *Applied Physics and Electrical Engineering Program* (Y-Program) initiated a study with the aim of inquiring into students' self-reported experiences related to:

- their expectations when they started the Y-program
- the curriculum and study environment in different phases of the program
- ways in which the program prepared them for their professional lives

The result was of great interest and value for further development of the curriculum, as well as for the improvement of the program. The project was later expanded to include groups of students from the 1998–1999, 1999–2000, 2000–2001, and 2001–2002 academic years.

The design of the study is based on:

- Questionnaires distributed to all students in each cohort, twice the first year, once a year during their study years 2 through 5, and one questionnaire a year after graduation
- Interviews with ten students, five male and five female, in each cohort. The same students are interviewed twice the first year, and once a year during their study years 2 through 5.
- Telephone interviews with the students one year after graduation
- Interviews with lecturers

Results from the study have continuously been fed back to the program board, and consequently, the Y-Program has improved in a number of ways, including taking action to attract and retain more students from underrepresented populations. Reports and conference papers have been published on the Y-Program website. When interpreting the results of these studies, it is important to note that outcomes of a study program cannot be predicted merely from the intentions of the program or the prerequisites and intentions of the students. It is a complex interplay between individual, institutional and political factors. In this longitudinal study, each new cohort enters with different experiences. The intentions of the program are, therefore, experienced differently. Results must be related to a context, in time and situation, as well as to reflections on the interactions between intended and unintended consequences of the actions taken.

– E. Edvardsson-Stiwne, Linköping University

Implicitly, the Initiative seeks to develop programs that are educationally effective *and* more exciting to students, attracting them to engineering, retaining them in the program and in the profession.

A standards-based approach to program evaluation provides evidence of a program's overall success in meeting these goals, and of the broader impact of our programs. To the extent that the standards measure inputs, processes, outcomes, and impact, program evaluation also measures improvement in these features. Considering the required development and implementation time from specification of mission to inputs, processes, and outcomes related to student learning, it is still relatively early to determine the overall impact of CDIO programs on their respective stakeholders. However, a few preliminary results are available.

Preliminary results of inputs, processes, and short-term outcomes

For several years, we have been collecting and analyzing data related to the key program evaluation questions posed earlier in this chapter. The use of multiple methods of data collection has yielded these preliminary evaluation results for inputs and processes:

- Self-reports of program evaluation status indicate that all participating programs have engaged in, or completed, the re-design of their respective curricula, the assessment of student skills, and the improvement of faculty competence.
- Self-reports and site visits indicate significant development and high degrees of student engagement in CDIO workspaces.
- Course evaluation results and instructor reflective memos indicate that faculty are using a wide variety of teaching and assessment methods.

We are using the evaluation tools to document steady progress toward full implementation of the inputs and processes standards. In addition, early indications show that programs are achieving their goal of improving the education of students, and the goals of attracting and retaining them in the profession.

- Self-reports indicate that all participating programs have adopted a mission that includes Conceiving-Designing-Implementing-Operating as the context and have engaged in stakeholder surveys to set learning outcomes.
- Annual surveys of graduating students indicate that they have developed intended program knowledge and skills.
- Student self-report data indicate high student satisfaction with the design-implement experiences and workspaces which promote a sense of community among learners.
- There is no evidence of any decrease in student knowledge or skills in technical fundamentals caused by, or as the result of, the integration of personal and interpersonal skills, and product, process, and system building skills.
- Longitudinal studies of students show increases in program enrollment, decreasing failing rates, particularly among students in underrepresented populations, and increased student satisfaction with their learning experiences.

Studies of long-term outcomes and overall impact

The 12 CDIO Standards focus on inputs, processes, and short-term outcomes. The Standards do not specifically address long-term impact, which is difficult to measure, and inherently involves long periods of time before meaningful data can be obtained. The evaluation of CDIO programs may also include questions related to students' future plans and employment after leaving the university. Such data can show attitudinal change and the beginnings of career behaviors that suggest long-term impact. In a much longer time frame, surveys reveal alumni contributions to their engineering fields, and the influence of a program on local, national, and international industries. It is too early to have meaningful program evaluation results of this kind.

A different dimension of impact is seen by the breadth of applicability of the CDIO approach. Program evaluation based on the Standards has recently been applied throughout Sweden to evaluate science, technology, and engineering programs at ten different universities. Box 9.2 describes the approach to program evaluation taken by the *Högskoleverket*, the Swedish National

Box 9.2. CDIO Program Evaluation and the Swedish National Agency for Higher Education (HSV)

The *Högskoleverket* (HSV), the Swedish National Agency for Higher Education, is the government agency responsible for the evaluation of university education in Sweden. Courses and professional degree programs are evaluated every six years. HSV also evaluates applications from universities and colleges to start new programs at bachelor, master, and doctoral levels.

In 2005, an evaluation of the *civilingenjör* engineering degree programs took place. These programs are 4.5-yr integrated engineering programs (to be 5-yr programs in 2007), roughly equivalent to Master of Science or *Diplome-Ingenieur* degrees. There are about 100 such programs in Sweden at about 10 universities. The programs range across all domains of science and engineering, including engineering physics, mechanical engineering, information technology, and industrial engineering. The questions are divided into university-level questions and program-level questions. During this evaluation process, HSV decided to add an overall program evaluation component to

- Complement the responses to the basic questions in order to attain a more comprehensive evaluation of the university and program
- Give the external review panel an additional instrument for its analysis and evaluation
- Provide the universities and programs with an instrument that could be applied as a basis for future continuous improvement efforts

The CDIO Standards and its associated rating scale were chosen for this purpose, with modifications to adapt the Standards to the context. When the self-evaluations were complete, program managers were surveyed to determine the usefulness of the CDIO Standards as a basis for program evaluation [7].

Survey and interview results indicate that the CDIO Standards are relevant and applicable to a wide range of programs, and that taking steps toward implementing the Standards would improve program quality. Survey results also indicated that the Standards' most important benefit is that they provide a basis for systematic program development. There were some concerns that the emphasis on personal and interpersonal skills, and product, process, and system building skills in the CDIO Standards might lessen the importance of disciplinary knowledge and engineering research. This concern can be addressed by supplementing the CDIO Standards with other key questions and instruments in an overall program evaluation. The full HSV report is available at their website.

– J. Malmqvist, K. Edström, S. Östlund, S. Gunnarsson

Agency for Higher Education [7]. The exercise and subsequent survey found the CDIO Standards to be broadly applicable and valuable in identifying pathways to program improvement.

As was acknowledged above, program evaluation based on the Standards may not address all the key questions needed for a complete picture of a program. As was found in the example in Sweden, other data collection activities may be required to supplement those related to the CDIO Standards.

SUMMARY

The CDIO Standards are useful in several ways for evaluating programs and curriculum change. They are based on the needs, goals, and approaches identified by CDIO programs and are founded on scholarship and emerging best practice. They provide a framework for the key questions focused on evaluation of program inputs, processes and outcomes. They can be applied flexibly to a wide range of programs, institutions, and academic cultures. The self-evaluation process results in specific actions for continuous program improvement and can be carried out on a regular basis, with few additional resources. Furthermore, the Standards can be used to guide new program development. They are not a generic program evaluation tool but emphasize the specific features of an engineering education set in the context of conceiving, designing, implementing and operating. The key questions and associated rating scale allow programs to evaluate current status, identify areas of potential improvement, plan reform, and benchmark their programs against peer programs worldwide.

The application of a standards-based approach to program evaluation is not without limitations, however. The key challenges to effective program evaluation are focused in two areas: implementing a variety of program evaluation methods to gather data from students, faculty, program leaders, alumni, and other key stakeholders that appropriately address the range of program outcomes; and documenting a continuous improvement process based on program evaluation results. Most engineering programs collect volumes of data about their students, faculty, facilities, and stakeholders. The challenge is to analyze the data and summarize results into information that is useful for decision making.

We have been using this standards-based program evaluation approach since October 2000. New collaborators conduct similar program evaluations as they begin their reform process, and as they project their desired status in two to five years. In Sweden, academic groups responsible for the evaluation of higher education programs have piloted the CDIO Standards as a component of their evaluation processes. The Standards are also consistent with evaluative criteria of national accreditation groups in the United States, Canada, the United Kingdom, and South Africa. With its emphasis on continuous program improvement, the standards-based approach augments accreditation reviews. At least annually, each program identifies specific tasks related to each standard to improve its overall program.

To this point in the book, we have examined, in detail, the key characteristics of CDIO programs, approaches to the design and development of a CDIO approach, and methods to implement and evaluate programs. We have seen examples from representative programs at each stage in their planning, implementation, and evaluation. In the final two chapters, we look back at the historical context of engineering education reform, and forward to anticipated changes in future engineering education programs.

DISCUSSION QUESTIONS

1. How can you use the CDIO Standards as the framework for evaluating your own programs?
2. What types of data or evidence do you rely on in your decisions about engineering programs?
3. How do you use evaluation results to improve curriculum, teaching and learning, student and instructor satisfaction, and learning spaces in your programs?
4. What would be the major impact of implementing a CDIO program in your institution?

References

[1] Worthen, B. R., Sanders, J. R., and Fitzpatrick, J. L., *Program Evaluation: Alternative Approaches and Practical Guidelines*, 2nd ed., Longman, New York, 1996.

[2] Kaplan, R. S., and Norton, D. P., *The Balanced Scorecard: Translating Strategy Into Action*, Harvard Business School Press, Cambridge, Massachusetts, 1996.

[3] Stake, R., *Standards-Based and Responsive Evaluation*, Sage Publications, Thousand Oaks, California, 2004.

[4] Weiss, C. H., *Evaluation*, 2d ed., Prentice-Hall, Upper Saddle River, New Jersey, 1998.

[5] Brodeur, D. R., *Report on the Self-Evaluation of Twelve CDIO Programs*, Unpublished report, 2006.

[6] Edvardsson Stiwne, E., *The First Year as Engineering Student—The Experiences of Four Cohorts of Engineering Students in Applied Physics and Electrical Engineering in Linköping University*. Paper presented at the 1st International CDIO Conference, Kingston, Ontario, June 2005.

[7] Malmqvist, J., Edström, K., Gunnarsson, S., Östlund, S., "Use of CDIO Standards in Swedish National Evaluation of Engineering Educational Programs, *Proceedings of the 1st Annual CDIO Conference*, Kingston, Ontario, June 2005.

CHAPTER TEN
HISTORICAL ACCOUNTS OF ENGINEERING EDUCATION

U. Jørgensen

INTRODUCTION

When engaging in the reform of engineering education, it is important to understand its historical context. For over 150 years, educational institutions have played a major role in shaping the skills and professional identities of engineers. During this period, the appropriate approach to engineering education has been the subject of constant discussions and controversy. Major changes have occurred both in the way engineering education is organized and in its relation to science education. Radical changes have also occurred in the technologies and technical specialties within engineering. Despite this history, and particularly in view of the controversies surrounding the role of engineering education since the late 1960s, engineering schools have been surprisingly stable in their basic philosophy regarding the structure and core content of the engineering curriculum. Only modest reforms have been implemented in the curriculum and pedagogy of engineering education in several decades. Most of these reforms have been focused on increasing the number of technical engineering topics, and solving the resulting problems of disciplinary congestion.

By the 1990s, organized efforts in both the United States and Europe raised basic questions about the relevance of engineering education as it had developed since World War II. The problems included a lack of practical skills in modern engineering training, the lack of relevance for industry of the science being taught, and the kind of analytical qualifications being awarded in engineering education compared with visions of engineers as creative designers and innovators of future technologies. With its emphasis on science and knowledge structured around technical disciplines, engineering education developed into an education of technically skilled cooperative workers. However, many feel that the knowledge and broad innovative capacity needed to produce creative design engineers able to cope with contemporary technological change seem to be lacking in engineering education.

Several educational initiatives have addressed these issues, and attempted to outline plans to reform engineering education. Some focus on engineering curriculum or pedagogy; some develop completely new engineering programs based on new technologies. Other initiatives combine business, management, and organizational understanding with engineering, or alternatively emphasize the creative and design aspects of engineering. Some reform initiatives have been supported by grants from government agencies, such as the National Science Foundation (NSF) in the United States; others have arisen from the Bologna Process that attempts to promote a unified system of education across Europe. While most initiatives focus on local, regional, or national experiments and reform, the CDIO Initiative is multinational, with open-source resources, and a broad, comprehensive methodology.

Contemporary tensions in engineering education may be deeply rooted in the diversity of modern technologies. The applications of these diverse technologies throughout society require increasing differentiation in the education of engineers. This diversity has already presented new challenges to the definition of engineering competence. The diversity of technologies presents new challenges to an engineering institution's sense of unity, identity, and standardization of professional preparation. Despite the complexity and multiplicity of technologies, institutional unity and its manifestation in a common engineering core curriculum have so far been successfully maintained by the engineering profession and by elite engineering universities. Nevertheless, the policies of identity formation and the creation of a homogeneous image of engineering are issues that need to be taken seriously, both in historical accounts and in contemporary reform initiatives. Engineering identity plays a vital part in educational reform and negotiations for change.

Critical accounts by observers close to the situation point to the need for reform in engineering education [1]–[2]. Other critics seem more confident in the achievements of engineers in society, and argue for the continuation of a traditional science-based engineering curriculum [3]. From their perspective, technology and the natural sciences are two distinctly separate approaches to knowledge [4]. Their studies contradict the popular misleading notion that engineering science is *applied science*. However, they do not raise critical issues related to the social and institutional dependencies of technology. Unfortunately, even engineering schools and professional institutions have supported the idea of a close relationship between science and technology by asserting that natural sciences are the core foundation of engineering. Contemporary developments in the natural sciences and engineering sciences have blurred the boundaries. New approaches of *techno-science* seem to be gaining ground as the characterization of the ties between modern science and technology, leaving neither one in a subsidiary role [5]. These new approaches recognize the role of technology as a contributor to scientific achievements, and change the basic idea of nature and technology.

The basic question is whether the critics are pointing to problems that will require radical reforms and transformations, or to a crisis in engineering education that will go away, as has happened so often before when technology and engineering have been criticized. The view that technology drives change and innovation seems to be less criticized today compared with the 1970s. At the same time, there is a crisis in engineering practice, itself, that relates to problems in the conception and use of technology, and in the needs expressed by industry and society pushing for reform.

The objective of this chapter is to set the CDIO Initiative in a historical context that traces the tensions of engineering practices, institutional changes, identity formation, and technological developments that are the context for modern engineering education. The intent is to highlight the complexities of the historical context, and not necessarily to produce a natural evolutionary history of engineering education that led to the development of the Initiative. The first section describes the early establishment of civil (non-military) engineering education, and illustrates models of engineering education that reflected diverse national identities and perceptions of the role of engineers in society. The next section delineates the role of engineering in industrial and societal development and how that role created the framework for the classical engineering specialties that led to international standardization of engineering subjects. The third section emphasizes the transformation to a *science* of engineering after World War II when more engineering science subjects were added to traditional natural science subjects. The discussions highlight the move away from practical skills in recent decades due to the diminishing need for skilled technicians and craftsmen who once formed an important recruitment base for engineering education. A description of the contemporary explosion in the number of new engineering disciplines, schools, and programs focusing on technological domains leads to a discussion of the controversy about what should be the engineering core curriculum of the future.

CHAPTER OBJECTIVES

This chapter is designed so that you can

- recognize historic changes and institutional differences in engineering education
- recognize the controversy over how engineering experience and practice should be represented in education
- understand the contribution of engineering education to the construction of engineering identity
- explain the reasons that reform initiatives have developed in recent years.
- evaluate the controversy between engineering problem solving approaches and the natural sciences as the foundation for an engineering core curriculum
- be inspired to experiment with new ways to provide students with engineering competencies

THE GENESIS OF ENGINEERING EDUCATION

Tensions between theory and practice have permeated engineering education since its formal inception in the 19th century. Scholars in the United States have used the metaphor of a swinging pendulum to describe various waves of practical orientation versus theoretical priorities setting the agenda for engineering education [6]. A closer study reveals a spectrum of positions, ranging from practical, skilled, craft-based technical education to science-based educations that developed in engineering schools and technical universities. The idea of the swinging pendulum is also seen in institutions with extreme polar differences in identities and focus, as was the case for a long time in several European countries.

Engineering professions emerged during the 19th century. Civil engineering developed first, as an offshoot of military engineering, which focused on the construction of armaments, fortifications, and infrastructure [3]. Early industrial work was based on practical skills and crafts, and led to the establishment of technical schools. Engineering, on the other hand, was based on a vision of technical development and the use of systematic, analytical approaches, similar to the French idea of *polytechnique* [7]. This idea was developed and promoted through the building of *École Polytechnique* in 1792, marking the beginning of a new era of civil engineering education. The ideas permeated both Europe and the United States in the first half of the 19th century and led to the establishment of a new type of institution of higher education. At the same time, military schools, such as West Point in the U.S., were heavily influenced by the analytical approaches that developed from the polytechnic idea. The practical and theoretical approaches led to distinct institutional structures of technical engineering and engineering education in Europe and the U.S. What today may seem to be a homogeneous profession with a well-defined international identity, has a hard-won and conflict-ridden history. We now take a brief look at the evolution of engineering education in France, northern Europe, the United Kingdom, and the United States.

Engineering education in France

In France, engineering institutions developed according to the structure of French government institutions and industry [8]. Inspired by the idea of *polytechnique*, the *grande écoles* have been the core of French state education, setting the ideal standards for education of engineers. Besides working in government institutions, engineers were involved in creating the new infrastructures demanded by growing cities and industries in need of transportation, energy, and communication. In this context, the technical sciences were seen as applied sciences. This thinking was based on the assumption that mathematical theory and general principles of science would form a foundation on which to improve technology moving it from the level of practice and skill-based experience to a higher form of practical knowledge.

Other engineering schools were established, several of which were focused on emerging sectors of industrial importance, such as the mining and mechanical industries which supplied agriculture and factories with new technology-based equipment. Though practical training was included and the demands of industry influenced the content of the curriculum, the elitist structure of these engineering schools maintained the hierarchy and roles of theoretical training.

Engineering education in northern Europe

In northern Europe, the dominant structure of engineering education consists of two models of engineering recruitment and education. One model, the *fachhochschulen,* is based on a practical education that recruits skilled craftsmen from industry and trades. This line of education developed in the late 19th century from technical schools to supplement the skills of workers coming from apprenticeship-based craft training by providing more theoretical subjects ranging from technical drawing to calculus [9]. The second model is a university-like academic engineering education, typically differentiated from the more discipline-oriented university education in natural science. Often named *technische hochschulen* (renamed technïsche universitätes in the late 20th century), a variety of technical universities developed in Germany and the Scandinavian countries to meet local institutional traditions. The basis for these two models was the strong tradition of skilled workers and the continued differentiation in identity, supported by two different recruitment paths: one offered to the practically skilled engineers; the other to the academically trained engineers coming directly from secondary school.

The second model gained legitimacy from the idea that the technical universities contributed to the production of reason, while the model of the *fachhochschulen* established its legitimacy by emphasizing its contribution to progress through its focus on quality techniques, the usefulness of practical engineering skills in industry, and the application of technology [10]. Engineers with academic educations contributed for more than 50 years to the construction of societal infrastructures and institutions [11]. Some of these theoretically trained engineers contributed to the rise of new industries following inventions in chemistry and electronics. However, in the 19th century, the numbers of practically skilled engineers still dominated industrial development in both the mechanical industries and in mining. Even in Germany where the theoretical training at engineering universities was initiated and supported by the creation of research and development facilities in larger corporations, the contribution of engineers in industrial innovation came from their practical experiences and systematic experiments, and only in small part from theoretical, science-based knowledge [12].

Engineering education in the United Kingdom

In the United Kingdom, a quite different institutional model developed. Engineering was seen as growing from the practical, skilled crafts and was therefore kept from the universities and the sciences. Although the idea of a polytechnic education found its way to the U.K. in the form of polytechnic institutions, its implementation resembled the class structure in society where leadership in government and industry was dominated by university graduates, and where engineering was seen as a secondary trade—important but based in practical skills. This division kept engineering education at a distance from the universities for quite some time.

In addition to the specific character of engineering education and the image of practical work in the U.K., the British system of accreditation formed an important difference between the systems created in Germany and France, which dominated continental Europe. In Europe, government committees defined the qualifications of engineers through their educational programs. The British system of accreditation emphasized practical skills and engineering experience, and it also supported the idea that engineering competencies were of a different nature than the academic qualifications given by universities. The U.K. system of accreditation has, to some extent, been copied in the United States.

Engineering education in the United States

In the United States, mechanical engineering was one of the first fields of engineering focus together with civil engineering. It emerged from the rich and diverse machine shop and agricultural machine cultures that sprang up to support industrialization. The earliest institution, Rensselaer Polytechnic Institute was founded in 1824 and acquired its modern name in 1861. Although its name resembles *polytechnique,* Rensselaer exemplified an American approach to engineering education that emphasized practical, industrial, and agricultural experiences for students, with comparatively less emphasis on mathematics and science. Other schools founded in succeeding decades emulated this essentially advanced apprenticeship model. In the mid-19th century, the establishment of the *Agricultural and Mechanics* (A&M) schools and *land grant* schools, including the Massachusetts Institute of Technology in 1861, reinforced this practical approach with close ties to industry, a dominant focus on practical knowledge, machine shop work, with little independent faculty research.

During the late 19th century, American engineering educators, for example, Robert Thurston, recognized the strengths of the European systems, and began to advocate for an increased presence of science and mathematics in the curriculum. This view coincided with an increasing desire for professional respect for engineering equal to fields such as medicine and law. Thurston's agenda also included adding a research emphasis. Many activities were included under this research umbrella, often through *engineering experiment stations* modeled on those in agriculture [6].

Although stirrings of new approaches appeared during the 1920s and 1930s, American engineering education remained largely within this practical, industrial orientation until World War II. In contrast, European schools, with such leaders as Felix Klein at Göttingen, excelled at applying scientific and theoretical approaches to engineering problems. During this period, intellectual leaders in the U.S., for example, Theodore von Kármán (a student of Klein), with European educations [6], transferred the idea of a more science-based type of engineering training to American institutions.

The practical approach of engineering institutions in the U.S. and in the polytechnics and *fachhochschulens* in Europe were of major importance to the development and implementation of technology in industry and society. These institutions influenced the formation of a professional engineering identity. Although this fact is recognized in contemporary discussions, it is overshadowed by the focus on the theoretical, science-based training that forms the modern ideal of formalized engineering teaching. The tension originates from the creation of an engineering identity where attempts to distance engineers from skilled technicians and their apprentice-based training resulted in a focus on an academic tradition based on the vision of the polytechnic institutions of higher education.

ENGINEERING AND INDUSTRIAL DEVELOPMENT

Many engineering universities and schools originate from civil and mechanical engineering developed during the first part of the 19th century. Their graduates were employed in government institutions, or involved in the creation of public and private companies that developed with the building of new infrastructures: transportation systems, roads, bridges, harbors, canals, ships, sewers, water supplies, and eventually, systems producing and distributing gas. Engineers' responsibilities and contributions to progress were based on their roles as constructors of the material pillars of modern society. Later, the view of their roles expanded to include engineers as innovators and system builders because of their contributions to new institutions, new knowledge, and the technical infrastructure [2]. To legitimize large investments in infrastructure, decision makers required hard data, and this need matched the focus on formal, science-based knowledge. The need to legitimize development also supported the creation of hierarchical and bureaucratic technological institutions. The idea of a technocracy that could support and even contribute to government policy was, therefore, consistent with the basic patterns and structures of knowledge created in relation to these large construction and infrastructure projects. For example, the connection in France between the idea of a *polytechnique* and the role of government bureaucracy is illustrated in the ideas of Hans Christian Oerstedt, the Danish founder of the *Polyteknisk Læreanstalt* (now the Technical University of Denmark) in 1829. Oerstedt saw a close relationship between polytechnic education and education in the political sciences in the German and Danish *staatswissenschaft* [13].

The role of military organizations, and the inspiration from large infrastructure projects employing military engineers also had a great influence on the factory systems established in large corporations [14]-[15]. These corporate systems were inspired by the military model of hierarchical organization and the need for unity and standards. The quest for standards also gave rise to the idea of improving productivity and sustaining control over the production process and work force through the use of scientific management principles, which soon became a core element of engineering management.

The tensions between the polytechnic training based on physics and mathematics versus practical skills in technical drawing and laboratory experiments were evident from the beginning. The controversies were fueled by the work of new polytechnic graduates, trained in the natural sciences, but primarily engaged in constructing the new technical infrastructures, that is, the water systems, sewers, gas pipes, and later, electrical power systems. The polytechnic graduates were also involved in establishing transport infrastructures, such as canals and rails, and the new communication infrastructures, such as telegraph, telephone and radio systems connecting cities, regions and nations.

In contrast, many of the machine shop developments and early industrial achievements were as much the result of the practically skilled technicians and craftsmen's work. They based their knowledge on the experiences of working with the construction of machines and chemical processes in industry, and they transferred their knowledge through visits to other sites. Traveling. working abroad, and returning with knowledge of technical constructions and innovations, detailed technical drawings and descriptions of new machines were common ways of transferring knowledge and new technologies [16]. The diffusion of the new technical constructs was supported by journals, most often national, such as the Danish *Polyteknisk Tidsskrift*, edited by Ursin, the first professor in mechanics at the *Polyteknisk Læreanstalt*.

During early industrial development, technical schools supplied many of the inventors of new machines, tools, and production systems. Practically skilled engineers, recruited from these schools, played an important part in industrialization until the late 19th century [11]. These engineers were involved in the new industries because of their experience with rationally organized experiments, documentation processes, and their experience with the achievements of developing technologies [17]. Though engineering disciplines show similar patterns across countries, differences of engineering practice exist across national cultures, especially in the way the theoretical contributions from engineering science are used in engineering practice [18]. In the later part of the 19th century, when research and innovations in petrochemicals gave a boost to the chemical industries and energy distribution systems, academic research and academically trained engineers gained importance. In this way, the developments of petrochemical and electrical technologies led to changes in the role of technical institutions of higher education, particularly in northern Europe.

The structure of many engineering institutions who built their curriculums on the *big four* in engineering – civil, mechanical, chemical and electrical – originates from this period. Although engineering schools still tended to train their students to solve practical industrial problems, and academic research was often difficult to distinguish from industrial consulting, electrical engineering was the exception. In this engineering discipline, the relationship between theoretical teaching and industrially developed technologies was closer than in other engineering domains. Still, many universities maintained basic engineering skills by requiring electrical engineers to study mechanics, technical drawing, and surveying. These requirements could not be explained in relation to the knowledge and skills needed in the new field of electrical engineering, but were established as part of the early curriculum for engineering education, in which mechanical and civil engineering practices were the standard.

In the course of history, many engineering disciplines developed from what could be called an *encyclopedia stage*, dominated by descriptive representations of technological exemplars, into a more abstracted and theory-based *scientific stage* [19]-[20]. This latter stage adds the strength of applying model descriptions, including mathematical representations and topic generalizations. However, in the transformation process, concrete experiences and practice-based knowledge, embedded in specific technical solutions, were often lost. Consequently, the transition represents a movement from scattered collections of representational exemplars to more complete representations of the technologies in question, documented by constructed theories and models. At the same time, the transition represents movement away from the engineering practice and experiences that are needed to make technology functional [21].

SCIENCE AS THE BASIS FOR ENGINEERING

In order to understand today's situation, we must consider one of the most important historical changes in engineering education – the construction of a science base for engineering. This development resulted from the increase in public and military funding of engineering research during World War II. The program to establish a science base for engineering created a new *elite* of theoretically oriented universities and technical schools of higher education in both the United States and Europe. At the outset, there was a gap in engineering curricula between science classes based on high degrees of mathematically formalized knowledge, and the more descriptive and less codified technical subjects. Earlier controversies resulted in positioning technical sciences as secondary, or applied, in relation to the natural sciences. Technical universities, at least in Europe, were restricted from giving doctoral degrees and addressing scientific matters without the support of university faculty versed in the natural sciences. However, the new era of expanding technical sciences lessened these controversies because of its increased focus on innovation and awareness of the close interactions between specific areas of science and technology.

Developments in the United States

The watershed event in American engineering was World War II. One of the leading institutions in this change was The Massachusetts Institute of Technology. Before the war, MIT had embraced scientific approaches, under the presidency of physicist Karl Compton. Vannevar Bush, a young MIT faculty member, reoriented his research from circuit simulations for electric power networks to general research in calculating machines – a more scientific orientation – successfully attracting private foundation support [22]. In 1940, Bush set up the National Defense Research Council, a major federal wartime research establishment in Washington, DC. Although engineers made significant contributions during the war, the success of the Manhattan Project put physicists in the spotlight, and savvy engineering leaders recognized that the path to prestige lay in engineers' closer emulation of scientists.

Developments in Europe

In Europe, this orientation toward a scientific basis for engineering already had a long tradition in the intellectual environment around the elite institutions, especially in France and Germany. The post-war tendency toward formalization of science councils and large government-sponsored research programs, centered on the peaceful utilization of technologies developed during World War II, spurred a dramatic increase in research at technical universities, and a change in the methods of teaching engineering. During the first half of the 19th century, several natural science subjects, for example in Germany and Denmark, were taught either in common at the universities and the polytechnic institutions, or only at the latter, so that natural sciences students had to take lectures at the polytechnic institutions. When the natural sciences became established within the traditional universities, they increasingly were perceived as being the foundation for the applied sciences.

During the first half of the 20th century, polytechnic universities had to fight for acceptance. They were acknowledged for their foundations in science, but were questioned about whether they could conduct independent scientific research; or were limited to practical experiments with technical improvements and practical implementation. These controversies manifested themselves in the acceptance of doctoral studies at technical schools of higher education. In Sweden and Germany, as in many other countries, decisions about what should qualify as scientific achievement and who was qualified to judge were very controversial. The controversy ended with an acceptance of technical or engineering science as a distinct area of scientific inquiry, although the image of engineering science as *merely* applied natural science continued to dominate many discussions about the character and role of technical sciences. Some engineering schools began granting degrees in engineering science to their best students [24].

Post-war developments

The movement toward a science base was concurrent with a massive post-war expansion of government-funded research in the United States. Sponsorship of fundamental studies in a variety of areas supported the trend away from practice-oriented research and education. Successes in fields such as high-speed aerodynamics, semiconductor electronics, and computing confirmed that physics and mathematics, conducted in a laboratory-based environment, could open new technological frontiers. Military research during these years also tended to focus on performance – increased power, higher altitudes, faster speeds – goals that were conducive to scientific approaches.

Electrical engineering, for example, no longer focused on electric power and rotating machinery, but instead, on electronics, communications theory, and computing machines. As historian Bruce Seely [6] wrote:

> *Theoretical studies counted for much more than practice-oriented testing projects; published papers and grants replaced patents and industrial experience as measures of good faculty. By the mid-1960s, the transition to an analytical and more scientific style was largely completed at most American engineering colleges."*

Yet today, many engineering departments still have their core activities defined by technical disciplines, such as mechanics, energy systems, electronics, chemistry, building construction, or sanitary and civil engineering. Many of these disciplines have specific problems and industries that relate to their founding years, but as the demand for science-based research and teaching became prominent, the original roots to practice and industry lost their significance. With the changing demands, more abstract courses, and courses defined by scientific fields, were developed.

The post-war decades saw the rise of systems engineering and thinking as broadly applicable engineering tools [22]. Systems sciences that include control theory, systems theory, systems engineering, operations research, systems dynamics, cybernetics and others led engineers to concentrate on building analytical models of small-scale and large-scale systems, often making use of the new tools provided by digital computers and simulations [25]. Techniques range from practical managerial tools, such as systems engineering, to technical formalisms, such as control theory, to more mathematical formulations, such as operations research. A broad-based movement within engineering found that these tools might finally provide the theoretical basis for all engineering that goes beyond the basic principles provided by the natural sciences. Whereas systems engineering of the 1950s could be narrowly analytical and hierarchically organized, new ideas of systems in the 1980s and 1990s focused on the relationship between technology and its social and industrial context. This new relationship and understanding of the natural and technical sciences is reflected in the notion that engineering as technoscience developed in the field of sociological studies of science and technology to reflect the new intimate relationship between these fields of science [26].

THE DECREASE IN PRACTICAL SKILLS AND EXPERIENCE

The creation of the research university as the ideal and elite model for engineering universities also influenced the staffing of engineering education's lecturer positions. The increase in research-based funding of these positions meant that the tradition of hiring practitioners to lecture in engineering was increasingly supplemented or replaced by lecturers hired on the basis of their achievements in engineering science and laboratory work, instead of their achievements in industrial practice. Voices were raised both inside and outside the universities against this change, resulting in the transformation of almost all lecturer and academic research staff at technical universities. However, within the universities, most objections and arguments came from practitioners who were involved in teaching and laboratory work identified as routine and trivial in comparison with frontline research.

The transformation of technical schools

In the European setting, the requirement of a Ph.D. degree narrows the recruitment of engineering faculty, making it difficult for engineers with careers in industry to satisfy entry requirements for university professorships. New Ph.D. recipients have been entering, in increasing numbers, into research positions funded by government programs, with fewer going to industrial laboratories and engineering practice. Even though the requirement of a Ph.D. degree can be substituted by personal innovative activities in industry, it is difficult to recruit practically skilled engineers to universities. Academic positions, today, require applicants to document research activities and demonstrate published works from their research in order to be evaluated for appointment. This threshold, in combination with a gap in the wage levels of industry compared with universities, has reduced the number of qualified engineers with skilled engineering practice in universities.

Changes in the foundation of engineering education, with the expansion of science-based technical disciplines, also led to changes in the curriculum of traditional vocational schools of engineering, as well as funding for research. Though with different names, the *polytechnics* in the United Kingdom, the *fachhochschulen* in Germany, and the *teknika* in Denmark shared common characteristics in recruiting students from groups of skilled technicians and supplementing their training with a theoretical education, while maintaining a focus on industrial practice. As a result, the schools inherited the experience-based, practical knowledge, and skills of students who had previously worked as apprentices in construction firms, machine shops, and industry. During the 1960s, the curriculum of these technical schools was expanded, and many of their specialized lines of engineering education were extended in length and scope. Typically, these changes included improvements in mathematics and natural sciences by copying the science base from engineering universities, while attempting to maintain their practical orientation. This led to the appointment of government committees to address the profile of practical engineering education [27]. It also raised questions about the balance between the academics and practice, and whether these schools would continue to supply practice-based engineers to industry.

At the same time, the decline in the apprenticeship training of craftsmen and skilled workers began to undermine the recruitment lines of the polytechnics [28]. While this type of engineering education was well supplied by the traditional, smaller crafts-based industries, the growth in the size of industries led to a change in the ways the workforce was trained, leading to an increasingly specialized machine shop skills in the workforce. Fewer candidates had the necessary broad skills and apprenticeship training required by the engineering schools. The schools were forced to establish other recruitment systems to survive. This process resulted in a complete reversal of the basis for recruiting students during the 1990s. As a result, it is difficult today to distinguish the two different lines of engineering education from one another, both because of the convergence of their student enrollments and the nature of their educational focus.

The response from industry

The response from industry to the tensions in technical education demonstrated the ambiguity of industry's interests in maintaining practically skilled engineers. Industry was not willing to carry the costs of an educational system to maintain the basic skills needed in the workforce. More generally, the problem also demonstrated the ambiguities in understanding which aspects of practice and experience were important for engineering work. Studies of engineering have demonstrated the importance of combining formal theoretical work, based on codified knowledge, with methods of drawing, experimentation, models, and analogous reasoning [16]. These skills cannot be based solely on the practical experience of shop-floor technicians, but also on the experiences of practicing engineers. Other practical perspectives, such as having experienced the daily routines of industrial organizations can be gained from other practices than working as an engineer. The ability of engineering institutions to recruit students with practical skills may have diminished, but the problem of maintaining the practical aspects of engineering competence continues to exist [29].

The return to practice

During the 1970s, a variety of technical and political events began to change the course of technology in its social context, and began to swing the pendulum yet again toward practice. The oil shocks, the beginning of the modern environmental movement, and the cancellation of the supersonic transport (SST) in the United States were indicators that technology might no longer progress along strictly technical lines. During the 1980s, the U.S. found itself in a crisis of competitiveness, which some blamed on the engineering research establishment's excessive focus on performance and military research, as opposed to other more industrial considerations. The *Made in America* study at MIT [30] reported that design and manufacturing had not received the academic resources or the intellectual prestige of the engineering sciences, and hence the U.S. had fallen behind such rivals as Germany and Japan in actually producing consumer goods. At the same time, the end of the cold war meant that large, military-oriented research funds might no longer be forthcoming. During the 1990s, academic institutions increasingly turned toward industrial sources of support. Along with new sources of money came new research orientations toward product design, product development, and innovation studies, with more emphasis on problems from engineering practice.

From within the technical universities, voices were raised against the consequences of a too-narrow focus on science-based teaching that lacked interest in the practical aspects of engineering work and competence [31]. Educational programs focusing on project work and problem-based learning, introduced in some experimental engineering education programs during the 1970s, spread broadly during the 1990s. They attempted to address the problems from a pedagogical and didactic point of view. In both Denmark and Germany, a few radical reform universities made project-oriented study the trademark of their education, stating that the projects could both cater to the interdisciplinary aspects of engineering methods and problem solving, and to the integration of the practical and theoretical elements needed in engineering [32].

While shop-floor training and practical aspects of work organization were the focus in the earlier phases of engineering, the new perspectives on engineering practice emphasize the complexity of engineering tasks, including project organization and communication, the role of specialized consultants, the skills needed to handle innovative design tasks, and the need to include the social dimensions of technology [33]. These new emphases may not eliminate the need for practical skills in drawing, visualization, modeling, and crafting of material objects, but the replication of traditional crafts does not satisfy the need for practical training in engineering. New emphases create a need to redefine *engineering practice* and to leave the apprenticeship model behind.

DISCIPLINARY CONGESTION AND BLURRING BOUNDARIES

The growth of the use of technology in the later half of the 20th century, in combination with the large investments made in engineering research by industry and by research institutes and universities, has resulted in tremendous growth in the body of technological knowledge, the number of new technological domains, and specialized technical science disciplines [34]. Differentiation in engineering specialties put pressure on engineering education to cope with the diversity and to keep up with the frontline of knowledge in the diverse fields. At many institutions, this resulted in a number of new specializations. Several of these specializations relate to sectors and industries that, for shorter or longer periods of time, have required engineers with particular kinds of knowledge. Changes in the demands for specialization created tension between generalized engineering knowledge and the specialized knowledge needed in individual domains of technology and engineering practice. Examples of these specializations include highway engineering, ship building, sanitary engineering, mining engineering, power generation and distribution engineering, offshore engineering, aeronautics, microcircuit engineering, environmental engineering, bioengineering, multimedia engineering, and wind turbine engineering.

Alternatives for addressing disciplinary congestion

All these specializations led to an expansion in the numbers and variety of courses focusing on technical sciences. At some technical universities, for example, the Massachusetts Institute of Technology and the Technical University of Denmark (DTU), the curriculum was organized into modules, giving students choices about how to structure their own education. While some universities expanded the number of specializations, others coped with disciplinary congestion through negotiation of the core content and opted for elective courses in only a limited part of the curriculum. General pedagogical reform based on project-oriented work also argued for giving students a broad understanding of engineering work and problem solving, with less emphasis on theoretical knowledge represented in the courses and disciplines. A very different response had been to question the concept of engineering education altogether, by giving more space in the curriculum to science-based teaching, by reducing the number of laboratory classes, and by weakening the ties to industry and the technological domains from which engineering originated.

Blurring boundaries between technology and nature

The dominant role of technology demands multidisciplinary approaches, and challenges the science-based, rational models and problem-solving approaches. These demands gave rise to new areas of engineering education. For example, in the field of environmental studies, the need for new approaches in industry based on cleaner technologies and product chain management challenged the already established disciplines in sanitary engineering based on end-of-pipe technologies and chemical analysis. From focusing on nature as a recipient of wastes, engineers had to realize that nature itself has been dramatically affected, and that environmental knowledge had to include the design of production processes and chemicals as part of what had become a continued re-design of nature. Blurring boundaries between technology and nature had introduced serious ethical and political issues into the core of engineering.

Another example can be found in the field of housing and building construction engineering. The need for integrating both social and aesthetic elements, as well as user interaction in both the project and use phases of construction, led to several attempts to overcome the traditional division between civil engineering and architecture. Several engineering education departments tried to solve this problem by employing staff from different disciplines – engineers, architects, and sociologists – hoping that solutions would emerge from the multidisciplinary melting pot. In several cases, the integration turned out difficult; housing construction and city planning in engineering crumbled in spite of the attempts. This dilemma left engineering housing construction departments in situations where the focus became more theoretical rather than contributing to the design and functionality of building construction. In contrast, functionality, usability, and flexibility, as well as the inclusion of users in the planning of building design, were left to the architects, who seemed more interested in aesthetics. This example illustrates the dominance of disciplinary culture in engineering schools, and the ways in which that culture defines and constructs new strands of knowledge and scientific research.

The influence of new technologies

Changes in the role of technologies in a society where consumer uses, complex production, and infrastructures are increasingly more important, have led to more focus on the integration of usability and design features. The traditional jobs in processing and production have not vanished, but new jobs in consulting, design, and marketing have been created. These new jobs demand new personal and professional competencies, and require new disciplines that contribute to the knowledge base [35]-[36]. During the 1990s, several engineering schools started new lines of education emphasizing engineering design skills and introduced aspects of social sciences into the curriculum of engineering design. These additions included technology studies, user ethnographies, and market analysis. The development of new and diverse technologies also reflects the limitations of technical sciences in being able to cover all aspects of engineering [31]-[37]. Examples of these reformed engineering programs can be found at Delft University in the Netherlands, Rensselaer Polytechnic Institute in the U.S., the Technical University of Denmark, the Norwegian University of Science and Technology, and Cranfield University in the U.K.

The growth and diversity of technological knowledge also leaves universities with continued pressure for renewal and difficulty in determining which engineering domains to maintain and develop. Several domains and branches of industry have passed their growth phase, and the related technologies are no longer the focus of research funding. The given industry may still be employing large numbers of engineers, but its need for new engineers does not justify creating or sustaining programs in engineering schools. For engineering education, it means the potential loss of important domains of technological knowledge.

The decade of the 1990s was not the first time that concerns about the role of technology in society had surfaced, but this time the questions raised issues of a more fundamental nature concerning the content of engineering education and the impact on technology exemplified with controversies about highway planning, chemicals in agriculture, nuclear power plants, and the social impacts of automation. The concerns questioned the role of knowledge about technology and some critics demanded a humanistic input into the curriculum with such subjects as ethics, history, philosophy, and disciplines from the social sciences [38]. This idea was based on the assumption that engineering students, through confrontation with alternate positions and opportunities to discuss social and ethical issues, would be better prepared to meet the challenges of technology. However, in most engineering education programs, these new subjects ended up being add-on disciplines often not integrated with engineering and science subjects, contributing further to the disciplinary congestion in engineering [39].

Developments in technology also have meant that the boundaries between engineering disciplines are blurring, and indeed the very nature and existence of engineering has come into question in recent years. What used to be fairly distinct areas of engineering – civil, mechanical, chemical, electrical – have now become combinations of two or more fields and their disciplines. For example, there are now programs in civil and environment engineering, aeronautics and astronautics, electrical engineering and computer science, and materials science and engineering. New programs in bioengineering and biomaterials reflect these shifts as well [1]. Today, many of the larger technical universities offer programs in more than a dozen different engineering fields.

Technological change has changed the face of engineering in many other ways, as well. Engineering research and design are changing, due, in no small part, to computers and the Internet. Algorithms, once taught as fundamental skills, are now built into automated design software. Large projects are run and coordinated through digital links between people who may never have met face to face. What had been largely a white male profession is now diverse in race, national origin, and gender. In addition, it is now possible for companies to conduct engineering functions worldwide with the help of automation and new technologies. The idea of well-defined boundaries for engineering education has been challenged by new technological domains and by existing university educations that already address technology as part of the curriculum.

Areas that address technology and have close affiliations with engineering represent a broad variety of subjects and approaches, for example, pharmaceuticals, architecture, computer science, information technology, environmental studies, biotechnology, nanotechnology, and technology management. These professional areas do not necessarily see themselves as part of engineering. In some areas, new perspectives of techno-science can create new relationships between science and technology. New fields of biotechnology and nanotechnology have blurred the boundaries with the natural sciences, as well, leading to the creation of such fields as mathematical engineering and nanotechnology in the natural sciences. Thus, a situation is developing where the new professionals, industrialists, and politicians question whether technology remains the domain solely of engineers, and whether engineering will continue to be the major source and producer of innovation. This development has been called *expansive disintegration* [1], reflecting the combined expansion of the number of technologies, specialties and disciplines on the one hand, and the continued disintegration of what once was the unity and identity of engineering on the other. These transformations will fundamentally challenge the role of engineering schools in the future.

CONTEMPORARY CHALLENGES

The role of engineers in technology and innovation is often taken for granted. Even in future-oriented reports on engineering, there is a tendency to expect problem solving abilities in societal and environmental issues from engineering, without challenging contemporary foundations of engineering curricula [23]. New insights coming from innovation theory, demonstrating a broader scope in innovation, coupled with changes in the societal use of technology that imply growing complexity and a need for social skills, point to the need for improvement in engineering education. On the other hand, innovations during the last decade are leading to changes that may make the role of engineering less central in the future. Policy and management attempts to govern innovation processes have also broadened the scope and shifted the focus from technological development and breakthroughs to a broader focus on market demands, strategic issues, and the use of technologies.

A new identity for engineering

Early in the 20th century, the idea that engineers have societal responsibility and are the heroic constructors of the material structures of modern society was being supplanted by a less heroic and more mundane image of engineers as the *servants* of industry. This image of engineering reflects a reduction in the influence of engineers on the direction and content of technological innovation, and supports the positioning of engineers in a less influential and subordinate role in their attempts to promote business interests [13]-[36].

This view is not all that different from engineers' self-image in contemporary society. The description of an engineer's competencies might include the following: possesses a scientific base of engineering knowledge, problem-solving capabilities, and the ability to adapt their knowledge and practices to new types of problems. The focus is more often on problem solving, and less on problem identification and definition [37]. This focus emphasizes the problem of engineering identity in distinguishing between engineers as creators and designers versus analysts and scientists. Although engineers' identity as creators and designers is supported in historical writing and in strategic reports about the role of engineering in the future [23], reality seems to place engineers in roles closer to analysts and scientists in laboratories and modern technical industries.

The underlying assumption in the discussion about engineering problem solving is that engineers are working with well-defined technical problems and methods from an existing number of engineering disciplines. This assumption does not answer the question as to whether engineers are competent in handling non-standardized social and technical processes where the problems are undefined and involve new ways of combining knowledge. Simply broadening the science base in a more interdisciplinary direction, including the social sciences and humanities, may not have been a satisfactory solution. The mere addition of topics to the curriculum does not change engineering practices or provide a better integration of knowledge [1]. A new engineering identity will be based on the answers to these questions:

- *What competencies are necessary to manage the creative, socio-technical and design skills that need to be improved in engineering education?*
- *What is the meaning of engineering problem identification and problem solving today, and how can they be reflected in engineering education?*

A new education for engineers

The reforms in engineering education, initiated in the 1970s in some engineering schools, emphasized the need for problem solving and project work that simulated real engineering practice, but these reforms did not provide the complete answer. The response lies in a new understanding of the role of science in innovation and the use of technology in context. This approach underlines the existing need to bridge the divide between the disciplinary knowledge of the technical sciences and social sciences, and the practical domains of engineering, with their unique knowledge and routines that integrate the social, practical, and technical aspects of technology at work [24]. It is necessary to rethink disciplinary knowledge as presented in engineering education, and a corresponding need to reform the content and structure of that knowledge.

One solution might be to accept that the idea of a single unifying engineering identity has proven to be problematic. Engineering education will unavoidably become more diverse in the future. Integrating engineering into the general university structure [1] could be a tempting solution, removing the rigid focus on core curriculum, while still fighting the battle for the acceptance of engineering science. However, the problems of including professional, practical knowledge and maintaining the need for professional skills in engineering are not solved by referring students to an even more diverse science base at universities. Neither do the many new science-based specializations in engineering provide a solution, and may even bring engineering further away from the practical knowledge also needed. Most technical disciplines focus on particular technical solutions taught as individual courses and with less emphasis on their application. These courses are supposed to contribute to a coherent set of engineering competencies, although they have little resemblance to an established domain of engineering practical problem solving and solutions [40].

Debates on engineering education tend to replicate a number of discussions over and over again. One example is the balance between practical skills and theoretical knowledge. While the debate may seem the same, the content has changed radically during the more than one century of controversy [41]-[42]. The list of relevant practical skills would not be the same, and similarly, theoretical knowledge has evolved as a result of technological developments, advanced tools, computers, and simulation models. Reforms need to produce a new realization of the kind of practical insights relevant to engineering education today.

Another unsettled challenge involves the balance between specialist and generalist knowledge in engineering. A process occurs in which the *current* deep knowledge and skills continuously changes. New knowledge and skills that begin as part of a science frontier are considered demanding. As the frontline of technological innovation moves, that knowledge and those skills become part of standard engineering procedures, technical standards, standardized components, design concepts, and are supported by computerized tools and simulation models. What counts as core or basic disciplines in an engineering curriculum changes in the wake of the expansion of new engineering domains and disciplines, despite the fact that all are dominated by the idea of a common theoretical foundation.

A new education for engineers will answer these questions:

- *What content should a core engineering curriculum have in the future?*
- *Which skills should be part of the curriculum, and which can be developed on the job after completing the education?*
- *What is the sequence of knowledge from abstract knowledge to practical application?*

Based on the insights from didactics, it is important to realize that the idea of a 'natural order of knowledge' starting in the most abstract and general disciplines and ending with application does not provide a good answer to curriculum planning. Instead, a critical approach to the role of knowledge in learning and the creation of engineering identity may be needed to overcome the taken-for-granted approaches in curriculum planning. Reform in engineering education needs to address the contemporary challenges highlighted here and find new ways of analyzing and understanding technological knowledge and the professional practices of engineering.

Addressing contemporary challenges with a CDIO approach

From the historical accounts presented in this chapter, we find a number of challenges that continue to be important in contemporary debates about the reform of engineering education. Key elements of the CDIO approach reflect critical issues underscored in this historical account. These elements include:

- A curriculum that includes personal and interpersonal skills, and product, process, and system building skills in the realization of technical solutions, while still emphasizing mathematics, the natural sciences, engineering science, and the technical knowledge specific to technological domains
- A core belief that personal and interpersonal skills, and product, process, and system building skills must be learned in the context of authentic problem solving and engineering practice
- An integrated approach to the teaching of engineering science and engineering disciplinary knowledge relevant to specific technologies and engineering practice

The CDIO Initiative renews the focus on several of the important issues in engineering educational reform that have been part of its history. While many reform initiatives have focused on the curriculum of different science and engineering disciplines, the balance between practical education and theory-based learning, or the role of project-based learning, the Initiative has created an approach that coordinates all the elements required to address the challenges of engineering in a modern, complex, technological society.

SUMMARY

In this chapter, we have shown that the engineering profession and engineering education have developed differently in European countries and the United States, demonstrating national differences in the role of engineering in society and industry and the ways engineering education has been influenced. The military use of infrastructure and equipment and the ingenuity of practically skilled builders pointed to the roots and inspiration for the creation of formal training of engineers for civil purposes and industrial development. Later developments emphasized the role of science in engineering training. The relationship between scientific theoretical knowledge in the disciplines, and the practical skills and knowledge derived from technological innovation and engineering practice has been controversial in engineering education and is still presents challenges for contemporary engineering education.

The differences among countries has led to the implementation of different educational structures characterized both by the background students were expected to have before entering the programs, and the ways in which lecturing was combined with laboratory and machine shop work. In the United States, with one dominant model of engineering education, the changing balance between practical and theoretical education has been characterized as a swinging pendulum, while in Europe, typically, two models of engineering institutions were created with differing weights placed on the two aspects of engineering.

With the rise of engineering science, especially since World War II, the core elements of engineering have been extended beyond the natural sciences. At the same time, the growth in the number of scientific specialties led to a focus on theoretical, science-based engineering education, leaving aside some of the creative design aspects of engineering, as well as some of the experiences and routines from the domains of engineering practice.

Engineering education has faced problems of a growing number of specializations with a consequent disciplinary congestion in the curriculum. At the same time, new areas of technology and new professional specializations outside engineering schools have introduced topics that are difficult to distinguish from engineering, blurring the boundaries between the professions. This blurring of boundaries, together with changes in the image of engineers away from *creator* toward *technical industrial worker*, have challenged the identity and content of engineering for the future.

The chapter concludes with questions about the knowledge base for engineering in modern society and its influence on a core curriculum for engineering education. This chapter points to a need to re-assess the composition of engineering education concerning the way practical skills and experiences are combined with theoretical training and the core elements of an engineering curriculum.

DISCUSSION QUESTIONS

1. Can you identify traces of the controversies in the history of engineering education, presented in this chapter, at your institution or in your professional society?
2. How would you characterize engineering education in your country? Can you identify one dominant type of engineering program in the way it balances theory and practice? Is there more than one model?
3. Do you see any limitations or barriers to an engineering program based solely on science and engineering disciplines? How are social issues of technology integrated into the curriculum?
4. In what ways can the CDIO approach change the trajectory of engineering education reform?

References

[1] Williams, R., *Retooling: A Historian Confronts Technological Change*, Cambridge, Massachusetts: MIT Press, 2003.

[2] Hughes, T. P., *Networks of Power: Electrification in Western Society*, Baltimore, Maryland: John Hopkins University Press, 1983.

[3] Auyang, S. Y., *Engineering: An Endless Frontier*, Cambridge, Massachusetts: Harvard University Press, 2004.

[4] Vincenti, W. G., *What Engineers Know and How They Know It: Analytical Studies from Aeronautical History,* Baltimore, Maryland: John Hopkins University Press, 1990.

[5] Ihde, D., and Selinger, E., *Chasing Technoscience: Matrix for Materiality*, Bloomington, Indiana: Indiana University Press, 2003.

[6] Reynolds, T. S., and Seely, B. E., "Striving for Balance: A Hundred Years of the American Society for Engineering Education", *Journal of Engineering Education*, Vol. 82, No. 3, 1993.

[7] Kranakis, E., *Constructing a Bridge: An Exploration of Engineering Culture, Design, and Research in Nineteenth-Century France and America*, Cambridge, Massachusetts: MIT Press, 1997.

[8] Crawford, S., "The Making of the French Engineer", in Meiksins, P., and Smith, C. (eds.), *Engineering Labour: Technical Workers in Comparative Perspective*, London: Verso, 1996.

[9] Gispen, K., *New Professions, Old Order: Engineers and German Society*, Cambridge: Cambridge University Press, 1990.

[10] Manegold, K., "Technology academized: Education and training of the engineer in the nineteenth century", in Layton, K. E., and Wiengard, P., *The Dynamics of Science and Technology: Sociology of the Sciences*, Dordrecht: D. Reidel Publishing, pp. 137-158, 1978.

[11] Nørregaard, G., *Teknikumuddannede ingeniørers betydning for den danske industri* (Engineers from Teknikum and their impact on Danish industry), Copenhagen: Ingeniør-Sammenslutningen, 1955.

[12] Hård, M., *Machines are Frozen Spirit: The Scientification of Refrigeration and Brewing in the 19th Century—A Weberian Interpretation*, Frankfurt, Germany: Campus, 1994.

[13] Wagner, M. F., *Det polytekniske gennembrud - Romantikkens teknologiske konstruktion 1780-1850 (The poly-technical breakthrough-The construction of technology in the romantic period 1780-1850)*, Aarhus: Aarhus Universitetsforlag, 1999.

[14] Roe-Smith, M., *Military Enterprise and Technological Change: Perspectives on the American Experience*, Cambridge, Massachusetts: MIT Press, 1989.

[15] Noble, D. F., *America by Design - Science, Technology and the Rise of Corporate Capitalism*, Oxford: Oxford University Press, 1977.

[16] Ferguson, E. S., *Engineering and the Mind's Eye*, Cambridge, Massachusetts: MIT Press, 1992.

[17] Hård, M., "The Grammar of Technology: German and French Diesel Engineering, 1920-1940", in *Technology and Culture*, Vol. 40, No. 1, 1999, pp. 26-46.

[18] Hård, M., *The Practice of Research: Behind the Scenes of Engineering Science*, paper forthcoming.

[19] Latour, B., *Science in Action - How to Follow Scientists and Engineers Through Society*, Cambridge, Massachusetts: Harvard University Press, 1987.
[20] Jørgensen, U., *Fremtidige profiler i ingeniørarbejde og -uddannelse, (Future profiles in engineering work and education)*, Copenhagen: IDA, 2003.
[21] Gibbons, M., Limoges, C., Nowotny, H., Schwarzman, S., Scott, P., and Trow, M., *The New Production of Knowledge - The Dynamics of Science and Research in Contemporary Societies*, London: Sage, 1994.
[22] Mindell, D., *Between Human and Machine - Feedback, Control, and Computing before Cybernetics*, Baltimore, Maryland: John Hopkins University Press, 2002.
[23] National Academy of Engineerng, *The Engineer of 2020: Visions of Engineering in the New Century*, Washington, DC: National Academy Press, 2004.
[24] Seely, B., "The Other Re-engineering of Engineering Education, 1900-1965, *Journal of Engineering Education*, July, pp. 285-294, 1999.
[25] Hughes, A. C., and Hughes, T. P., *Systems, Experts, and Computers: The Systems Approach in Management and Engineering, World War II and After*, Cambridge, Massachusetts: MIT Press, 2000.
[26] Juhlin, O., and Elam, M., "What the New History of Technological Knowledge Knows and How It Knows It", in Juhlin, O., *Prometheus at the Wheel: Representations of Road Transport Informatics*, Linköping, Sweden: Tema T, Linköping Universitet, 1997.
[27] Finniston, M., *Engineering Our Future. Report of the Committee of Inquiry into the Engineering Profession*, London: Her Majesty's Stationery Office, 1980.
[28] Lutz, B., and Kammerer, G., *Das Ende des graduierten Ingenieurs? (The end of the 'craft-based' engineer?)*. Frankfurt: Europäische Verlagsanstalt, 1975.
[29] Cohen, S. S., and Zysman, J., *Manufacturing Matters - The Myth of the Post-Industrial Economy*, New York: Basic Books, 1987.
[30] Dertouzos, M. L., Lester, R. K., and Solow, R. M., *Made in America–Regaining the Productive Edge*, Cambridge, Massachusetts: MIT Press, 1989.
[31] Seely, B., "A Swinging Pendulum: The Place of Science in American Engineering Schools, 1800-2000", in Jørgensen, U. (ed), *Engineering Profession and Foundations of Technological Competence*. (forthcoming)
[32] Kjersdam, F., and Enemark, S., *The Aalborg Experiment -Implementation of Problem Based Learning*, Aalborg: Aalborg University Press, 2002.
[33] Henderson, K., *On Line and On Paper: Visual Representations, Visual Culture, and Computer Graphics in Design Engineering*, Cambridge, Massachusetts: MIT Press, 1999.
[34] Wengenroth, U., *Managing Engineering Complexity: A Historical Perspective*, paper for the Engineering Systems Symposium at MIT, 2004.
[35] Sørensen, K. H., "Engineers Transformed: From Managers of Technology to Technology Consultants", in *The Spectre of Participation*, Oslo: Scandinavian University Press, 1998.
[36] Schön, D. A., *The Reflective Practitioner: How Professionals Think in Action*, New York: Basic Books, 1983.
[37] Bucciarelli, L. L., *Designing Engineers*, Cambridge, Massachusetts: MIT Press, 1996.
[38] Beder, S., *The New Engineer: Management and Professional Responsibility in a Changing World*, The University of Wollongong, 1998.

[39] Downey, G., "Are Engineers Losing Control of Technology? From 'Problem Solving' to 'Problem Definition and Solution' in Engineering Education", *Chemical Engineering Research and Design*, Vol. 83, 2005.

[40] Knorr Cetina, K., *Epistemic Cultures - How the Sciences Make Knowledge*, Cambridge, Massachusetts: Harvard University Press, 1999.

[41] Björck, I. (ed.), *Vad är en ingenjør? (What is an engineer?)*, report from the NyIng project, Linköping, Sweden: Linköpings Tekniska Högskola, 1998.

[42] ATV, *Ingeniørernes nye virkelighed–roller og uddannelse*, (The engineers new reality, roles and education). Lyngby, Denmark: Akademiet for de Tekniske Videnskaber, 2000.

CHAPTER ELEVEN
OUTLOOK

WITH S. GUNNARSSON

INTRODUCTION

The CDIO Initiative responds in an integrated and pragmatic way to the historical context in which engineering education finds itself and to the challenges that lie in the future. The Initiative began with four universities in two countries, and has expanded rapidly in terms of scope and participating universities. The initial programs were typically within the domains of mechanical, vehicular and electronic engineering, but the CDIO approach has now been implemented in programs in chemical engineering, material science and engineering, and bioengineering. The model has been applied to reform initiatives affecting all engineering programs at a university, and as a template for national initiatives and evaluation schemes. The number of universities has now expanded to more than 22 universities in 12 countries on nearly every continent. Development is underway at universities characterized as research-intensive or teaching-focused; large or small; private or public; or historically focused on minority and underrepresented populations. Regional CDIO Centers in North America, the Nordic countries, the United Kingdom and Ireland, and Southern Africa, have been established to provide opportunities for the exchange of ideas and support for implementation in local regions. A number of vehicles, tools, and forums for disseminating and developing the CDIO approach have been created, including the website and the annual international conferences.

The CDIO approach is likely to evolve and be adapted and implemented in an even wider variety of settings—in engineering disciplines not already covered, in graduate education, and in education beyond engineering. It has been designed to be flexible and adaptable, with the ability to respond to the forces driving engineering education in the near future. We look forward to working with others in this evolutionary process. This chapter highlights what we see as future challenges for engineering education, and outlines ways in which a CDIO approach can be developed to address these challenges.

CHAPTER OBJECTIVES

This chapter is designed so that you can

- recognize the factors that continue to drive change in engineering education and ways in which the CDIO Initiative relates to them
- discuss the potential for development and broader application of the CDIO approach

DRIVERS FOR CHANGE IN ENGINEERING EDUCATION

The major goal of engineering education is to serve society and engineering students by providing up-to-date, high-quality learning opportunities, organized as programs. Maintaining and improving quality requires an awareness of the key environmental factors that drive change in engineering education. The most important drivers for change in engineering education include:

- Scientific breakthroughs and technological developments
- Internationalization, student mobility and flexibility
- Skills and attitudes of beginning engineering students
- Issues of gender and broadening participation
- Governmental and multilateral policies and initiatives

It is important to have good mechanisms in place for maintaining awareness of the factors that drive change and to have effective methods to plan and implement changes in engineering programs. Chapter Eight suggests methods for implementing program change, and Chapter Nine gives examples of tools and techniques for program evaluation and improvement. The CDIO Syllabus, itself, discussed in Chapter Three, can also be a useful tool for monitoring some of the drivers.

Scientific breakthroughs and technological developments

Scientific and technological evolution is an obvious driver for the development and improvement of engineering education. Existing subjects in the curriculum have to be updated according to the progress within the discipline, and new fields of study need to be incorporated into the curriculum. There are several ways to keep a curriculum current and relevant. One way is to see to it that faculty have sufficient resources for research within their disciplines. Relevant research results can then be introduced into the educational program. A second way is to assure that adequate mechanisms exist for bringing developments in industry into the engineering education program. We can achieve closer ties with industry by hiring faculty and research staff with industrial experience and by involving people from industry in program implementation and management.

Technological developments also influence engineering programs in the way that design, development, and production are organized and geographically located. For many industrialized countries today, manufacturing and production are moving outside their borders to other countries where production and labor costs are lower. If engineering research, design, and development follow the export of production and manufacturing, the change will substantially influence the need for engineers and their expertise, and consequently, engineering education.

A CDIO approach offers several ways to keep the education up to date with changes in science and technology. As described in Chapter Six, CDIO Standard 7 *Integrated Learning Experiences* emphasizes real-world problems in engineering education through involvement of industrial partners in the formulation of learning experiences. The Syllabus, described in Chapter Two, is another tool for tracking the development and needs of industry. Results of stakeholder surveys, especially answers from people active in industry, are obvious input to the process of educational development. Finally, faculty who are involved in both education and research are better able to influence engineering programs as a result of scientific breakthroughs and technological development. As explained in Chapter Eight, Standard 9 *Enhancement of Faculty Skills Competence* encourages this latter kind of involvement.

Internationalization, student mobility and flexibility

The globalization of current workplaces and companies requires that graduates be prepared for careers characterized by daily international contacts, frequent travel, and extended distance collaboration. It follows that education will become increasingly international and that it must lead to internationally recognized degrees. We can see that already in the tremendous change in the mobility of students that has taken place during the last two decades. In Europe, for example, student mobility increased with the creation of European student exchange networks, such as *Erasmus* and *Socrates*. Mobility and flexibility are also important aspects of the Bologna Process, because a uniform structure of higher education increases students' opportunities to move between universities. Box 11.1 is a description of key points of the Bologna Process [1] as applied to Sweden and the United Kingdom.

In North America, there is likewise a long and growing tradition of international mobility in education. For example, in Canada, the largest numbers of international students come from China, India and the Middle East. They usually complete a four-year Bachelor of Science program, which is typical for all engineering programs in Canada, consistent with the accreditation requirements of the Canadian Engineering Accreditation Board (CEAB) [2]. In the United States, there are increasing numbers of programs for engineering students to study abroad for at least one year. Throughout the world, national higher education systems, including those in Chile and Australia, are considering large-scale structural changes to allow greater mobility of their students, as well.

Box 11.1. The Bologna Process in Sweden and the United Kingdom

The Bologna Process is a joint European effort, involving 40 countries, to obtain a uniform structure for higher education in the European countries. The Bologna Declaration involves six actions relating to higher education:

- A system of academic grades that are easy to read and compare
- A system essentially based on two cycles
- A system of accumulation and transfer of credits
- Mobility of students, teachers, and researchers
- Cooperation with regard to quality assurance
- The European dimension of higher education

The aim of the process is to make the higher education systems in Europe converge toward a more transparent structure whereby different national systems would have a common framework based on three cycles - Bachelor, Master and Doctorate.

There is not yet an established agreement on what should be the objectives for the three levels. One proposal is the so-called *Dublin Descriptors* [1] that state that
Bachelor degrees are awarded to students who:

- have demonstrated knowledge and understanding in a field of study that builds upon and supersedes their general secondary education and is typically at a level that, while primarily supported by advanced textbooks, includes some aspects that will be informed by knowledge of the forefront of their field of study
- can apply their knowledge and understanding in a manner that indicates a professional approach to their work or vocation and have competencies typically demonstrated through devising and sustaining arguments and solving problems within their field of study
- have the ability to gather and interpret relevant data (usually within their field of study) to inform judgments that include reflection on relevant social, scientific, or ethical issues
- can communicate information, ideas, problems, and solutions to both specialist and non-specialist audiences
- have developed those learning skills that are necessary for them to continue to undertake further study with a high degree of autonomy

Master degrees are awarded to students who:

- have demonstrated knowledge and understanding that is founded upon and extends and/or enhances competence typically associated with the Bachelor level, and that provides a basis or opportunity for originality in developing and/or applying ideas, often within a research context
- can apply their knowledge and understanding, and their problem solving abilities in new or unfamiliar environments within broader (or multidisciplinary) contexts related to their field of study
- have the ability to integrate knowledge and handle complexity, and to formulate judgments with incomplete or limited information, but that include reflecting on social and ethical responsibilities linked to the application of their knowledge and judgments
- can communicate their conclusions, and the knowledge and rationale underpinning them, to specialist and non-specialist audiences clearly and unambiguously
- have the learning skills to allow them to continue to study in a manner that maybe largely self-directed or autonomous

Doctoral degrees are awarded to students who:

- have demonstrated a systematic understanding of a field of study and mastery of the skills and methods of research associated with that field
- have demonstrated the ability to conceive, design, implement, and adapt a substantial process of research with scholarly integrity
- have made a contribution through original research that extends the frontier of knowledge by developing a substantial body of work, some of which merits national or international refereed publication
- are capable of critical analysis, evaluation, and synthesis of new and complex ideas
- can communicate with their peers, the larger scholarly community, and with society in general about their areas of expertise
- can be expected to promote, within academic and professional contexts, technological, social, or cultural advancement in a knowledge-based society

The effort to adapt engineering education in a specific country to the Bologna structure depends on the national goals and organization prior the Bologna Process. For example, in Sweden, engineering education consisted of three-year programs leading to a Bachelor degree (*Högskoleingenjör*), and 5-year programs leading to a Master degree (*Civilingenjör*). The degree *Civilingenjör* has a long tradition, and represents a strong brand in Sweden. Therefore, the government proposed that this degree continue to exist after the introduction of the three-cycle system. The main challenge for Swedish universities is to find suitable forms of co-existence between the three-cycle system and engineering programs leading to the degree *Civilingenjör*.

---

[1] Joint Quality Initiative Group, *Shared 'Dublin' Descriptors for the Bachelor, Master and Doctoral Awards*, Dublin, 2004.

– S. Gunnarsson, Linköping University and J. Malmqvist, Chalmers University of Technology

In the United Kingdom, the Master of Engineering (M.Eng.) degree has an equivalently well-established brand image, despite the potential for confusion inherent in its "Master" name. The four-year M. Eng. degree was considered in the United Kingdom to be an undergraduate first-cycle degree, with enhanced content and greater breadth than the conventional 3-year Bachelor of Engineering. Rather than a Bachelor degree with an optional add-on year, a 3+1, it was seen as 4+0. However, there are now suggestions that it should be viewed as an *integrated Master* second-cycle degree. The M.Eng. already co-exists with a postgraduate, that is, second-cycle Master qualification, the Master of Science, which usually takes 12 months rather than the typical 24 months of a second-cycle qualification elsewhere and for which the entry qualification is usually a three-year Bachelor of Engineering or Bachelor of Science. A challenge for institutions in the United Kingdom is to reconcile these three degrees (B.Eng., M.Eng. and M.Sc.) with the two-cycle of the Bologna pattern prior to the Doctorate. It is often considered an advantage that the system in the U. K. can take a graduate to the end of the second cycle within four years, and there is resistance to any move toward a 3+2 model.

– P. Goodhew, University of Liverpool

Commonality in accreditation is another mechanism of internationalization. International agreements, such as the *Washington Accord* [3] on cross-recognition of professional certification cause various accreditation schemes to converge. In the United States, the accreditation criteria of the American Board of Engineering and Technology [4] have influenced thinking in many national systems, and attracted many international programs to apply for accreditation. In Europe, there is an initiative, aligned with the Bologna Process, that is developing a common system for accreditation of engineering education. The project, *Accreditation of European Engineering Programmes and Graduates* (EUR-ACE) [5], has been officially accepted by the European Commission. The aim of the project is to create an accreditation system that is compatible with the systems currently used in certain European countries [6].

The CDIO Initiative supports internationalization and mobility by providing a well-developed international model, a basis of common comparison of student learning outcomes, and potentially the basis for common accreditation. Meeting the Bologna Process and accreditation criteria will be a basic requirement of all educational programs in the future. However, accreditation requirements are high-level and formal in character. The Initiative takes a further step toward a truly international education by implementing and adapting a pragmatic model, developed in collaboration by leading universities around the world. Within CDIO programs, there is a close connection between accreditation and the learning outcomes of the Syllabus. In Chapter Three, we compared the CDIO Syllabus with ABET's evaluative criteria, specifically EC 2000 Criterion 3. (See Table 3.3) The national evaluation of engineering education carried out in Sweden during 2005 is another example of connections between a CDIO approach and national accreditation and evaluation efforts. The Swedish Agency for Higher Education (HSV) used the CDIO Standards as a core component of the self-evaluation completed by all universities in Sweden offering engineering education. (See Box 9.2 in Chapter 9)

Skills and attitudes of beginning engineering students

The skills and attitudes of students entering engineering programs are important drivers for the ways that education is designed, both in terms of content and organization. Education systems are part of the surrounding society; hence, changes in societal attitudes affect engineering education. Many industrialized countries are experiencing decreased interest in science and technology among younger students. This lack of interest influences engineering education in that fewer, and less motivated, students pursue engineering. Attitudes toward science and technology in a society also affect the importance placed on these subjects in secondary education.

In addition, universities in many countries face increasing difficulties with the level of knowledge and background experience of entering students. This is a recognized fact in mathematics and physics [7]. It is also vital that engineering education address the development of the practical skills and technical knowledge gained through pre-college curriculum activities and life experiences, such as tinkering with electronics, building things, repairing everyday devices, and developing software. Such pre-college experiences, more common in the past, facilitate the acquisition of theoretical knowledge, by connecting it to practice.

Addressing these issues requires changes on many levels in the school systems – primary, secondary and university levels. Within the university, introductory courses aim to orient students to the role of science and technology in society and provide initial engineering experiences that strengthen students' motivation. Hands-on and design-implement learning activities provide concrete experiences that connect abstract models for mathematics and physics with practical applications. Such experiences also explicitly seek to make engineering more interesting and exciting, recruiting students to engineering and retaining them in the profession. Design-implement experiences are being considered as extensions to the curriculum in primary and secondary schools, further strengthening students' motivation and preparation to study engineering at the university level.

Issues of gender and broadening participation

Throughout the world, there is significant interest on the part of educators and government to increase the participation in engineering of women and populations that have been historically underrepresented or disenfranchised. Engineering is viewed as a profession of upward mobility, which has the potential to positively influence the well-being of society. For these reasons, nations have an interest in making engineering education accessible to qualified students, regardless of their backgrounds.

The CDIO Initiative has been supportive of this effort. We have studied, for example, how gender and related issues manifest themselves in our educational programs. This has highlighted the need for better role models. In addition, our experiences show that it is important to choose examples, project tasks, and other learning experiences that appeal to a broad segment of the student population.

In many countries, there are ongoing discussions about how to influence the attitudes of all young people toward engineering. Students' attitudes toward engineering education are influenced by several internal and external factors. The structure, content, and organization of the engineering education itself are important factors. To date, evidence gained at two universities show that women and underrepresented minority students who participate in first-year design-implement courses are more likely to complete their engineering programs.

Governmental and multilateral initiatives

The development of engineering education programs takes place on several levels, from individual faculty levels to national and international levels. Decisions taken at higher levels create boundaries and conditions for development at lower levels. In Europe, the Bologna Process, described in Box 11.1, is a good example of a multilateral initiative. Once multilateral agreements are made at the European level, each participating university interprets policies and makes decisions for its own educational system. Principles defined at the national level then become the starting point for each university in developing its programs. When an individual university has formulated its strategic plans, program development reaches specific engineering programs and courses.

The Initiative supports this coordination and planning in several ways. Program development focuses on the last two levels, namely, the program level and the course level. The flexibility inherent in our non-prescriptive resources allows tailoring to local and disciplinary contexts. The commonality of the CDIO approach facilitates international benchmarking and collaboration.

FUTURE DEVELOPMENT OF THE CDIO APPROACH

The CDIO Initiative represents a collaborative effort to reform engineering education. As described in Chapter Two, a number of universities around the world are involved in partnerships and consortia to improve engineering education. Table 2.1 in Chapter Two lists five examples of such consortia. We acknowledge and admire these and similar efforts aimed at broad-based reform. We hope to influence them and learn from them. Unlike some reform projects, the Initiative is not primarily focused on educational research. Programs apply ideas and adopt methods that have been shown to be a part of the best practices in science and engineering education and the outgrowth of scholarly research on education. We document the design and implementation of curriculum reform efforts and share these with other engineering educators. Our ambition is to continue to develop the CDIO approach by working with other partnerships, as well as individual reformers and researchers,. We hope to disseminate our findings widely. In this section, we discuss the potential for applying the CDIO approach in additional engineering disciplines, in graduate programs, and in fields beyond engineering.

Application to additional engineering disciplines

The first collaborators came from the engineering disciplines of mechanical, vehicular, aerospace, and electrical engineering. These disciplines are distinguished by discrete serial production products that are systems. The examples, terminology, and thinking are somewhat biased by these origins. However, in order to show the ability to generalize the CDIO approach, it is important that the approach be tried in additional traditional engineering disciplines, such as civil and chemical engineering, as well as emerging engineering fields, such as bioengineering and nanoengineering. This dissemination is the focus of both existing and new collaborators.

The aim of applying Conceive-Design-Implement-Operate to other engineering areas brings out a need to answer key questions:

- *Can Standard 1, the product, process, and system lifecycle context, be generalized? Is it applicable to other disciplines?*
- *Are there pedagogical and curricular differences in applying a CDIO approach to:*

  - *Other traditional engineering disciplines, for example, civil, ocean, software engineering*
  - *Fundamental science and engineering disciplines, for example, material science, bioengineering, nanoengineering, applied physics*
  - *Industrial engineering, manufacturing engineering, and engineering management*
- *Can a program adapt the approach in part, and if so, what percentage of the Standards must be incorporated in order to be considered a "CDIO Program"?*

Generalizing the product, process, and system lifecycle context. Standard 1, *the adoption of the principle that product, process, and system lifecycle development and deployment—Conceiving, Designing, Implementing, and Operating—are the context for engineering education,* may seem very closely tied to the initial disciplines of the CDIO Initiative. The terms, *systems*, *products*, and *implementation*, may not feel comfortable for a program in civil or chemical engineering. While an academic program manager may be able to translate the terminology to that of the domain, other stakeholders may not do so willingly. However, it is perfectly feasible to change the terms used in Standard 1 to fit a particular field of engineering while still keeping the intent of the standard intact, by focusing on what is designed and implemented by the engineer. For example, civil engineers are likely to prefer to speak about *buildings* rather than *products*, and an adapted version of Standard 1 could read *"The principle is to educate engineers to meet the needs of the construction industry, that is, planning, design, engineering, production, operations and maintenance of buildings"* [8].

Other changes in terminology may follow, once the decision is made to adapt the wording of Standard 1 to a particular domain's context. These may include terminology in other standards, and in Section 4 of the CDIO Syllabus—*Conceiving, Designing, Implementing and Operating Systems and Products in an Enterprise and Societal Context.*

Pedagogical and curricular differences. More substantive changes than terminology may be necessary when adapting the CDIO approach to disciplines in which the nature of the design-implement sequence is fundamentally different from the development of discrete products or systems. In bioengineering, for example, the design and implement process is not easily described by end goals, but more aptly as reaching the limits allowed by physics, chemistry, and biology. It may not be possible to decompose the overall problems into separately solvable problems that can then be integrated into a system solution. Indeed, the Engineering Biology Program at Linköping University identified the interpretation of the design-implement concept as one of their key challenges. Box 11.2 is a brief description of their Engineering Biology Program.

There are creative ways to incorporate design-implement experiences in biological engineering. Molecule-level variants of design-implement learning experiences may be developed, for example, by using site-directed mutagenesis to modify the function of a specific protein in a microorganism [9]. Such a

Box 11.2. Conceiving-Designing-Implementing-Operating in Engineering Biology at Linköping University

The Engineering Biology Program at Linköping University (LiU) started in 1996. The program is 5 years, with the first three years focused on mathematics, physics, chemistry, biology, and engineering. Engineering courses include programming, electronics, automatic control, and signal processing, among others. The fourth year is devoted mainly to a specialization. Eight specializations are currently available, including bio-informatics, micro-systems and bio-sensors, and protein engineering.

During 2004, the Engineering Biology Program Board formulated a plan to strengthen the engineering aspects of the program. The CDIO model is an essential component in the plan. The first step in the transformation to a CDIO program is an introductory course that was offered for the first time during 2005. Work has also begun on the development of program and course learning outcomes. Project courses, connected to the different specializations, will be introduced later in the program.

One key issue in the introductory course, and subsequent project courses, is how to address the design-implement concept. The interpretation of *product, process, and system* in the CDIO Syllabus and the CDIO Standards needs careful consideration. In the first version of the introductory course, several projects will deal with the design and implementation of systems for measurement and monitoring of biological processes, applications that are close to the disciplines of the original CDIO programs but still within the scope of the Engineering Biology program.

– S. Gunnarsson, Linköping University

learning experience might start with the students designing a modified gene sequence and predicting the consequences on the protein structure. The next step is to produce a plasmid containing the modified gene, and transfect a bacterium with the plasmid. The bacterium is then cultivated to produce the recombinant protein. Finally, the function of the protein, or of the genetically modified bacterium, is evaluated using biochemical methods. Learning experiences can be enhanced by having students keep laboratory notebooks, in which they document all processes.

Other programs are adopting a CDIO approach to bioengineering as well. The mechanical and materials engineering program at Queen's University in Canada will introduce a second option in biomedical engineering in the Fall of 2007. At the University of Liverpool, the approach has been adopted in a program that includes material science and engineering, as described in Box 11.3.

Adapting and adopting parts of the CDIO approach. The programs involved in the Initiative have the stated aim of implementing all twelve CDIO Standards. However, programs may find some parts useful and others less relevant or unrealistic in their circumstances. This raises the question of what percentage of the Standards must be incorporated to be considered a "CDIO Program"? There is no distinct threshold where a program becomes, or ceases to be, a CDIO program. However, it would be hard to imagine a CDIO program that did not accept some variant of Standard 1, acknowledging that the product,

Box 11.3. Conceiving-Designing-Implementing-Operating in Materials Programs at the University of Liverpool

Three-year Bachelor of Engineering and four-year Master of Engineering programs in materials science at Liverpool are being re-cast to comply fully with the CDIO Standards. They already share 94% of a common first year with other engineering programs (Mechanical, Aerospace, Civil, Product Design). Each first-year program also has a small differentiating module designed to introduce the flavor of the sub-discipline. For Materials students, this involves teamwork to develop a classification scheme for materials. This scheme has most of the attributes of a *product* and certainly requires a systems approach.

All first-year students undertake two design-build-test exercises in teams of five or six. Thus, overall compliance with CDIO objectives is very high. In the second, third, and fourth years, there are a number of ways in which Materials students benefit from operating in a broad engineering department. They are ideal team members able to contribute materials selection input to CDIO exercises, and they integrate particularly well with Product Design majors. An example of a problem-based module taken by Materials students is the "car door" exercise. Teams of students are tasked with improving the performance (weight, dent resistance, and cost) of an existing steel car door design. They necessarily have to engage with product design, materials selection, testing (both real and virtual, using specially designed software), and interpersonal skills such as reporting to company personnel and negotiating advice. So far all of these activities have been well received by the students.

– P. Goodhew, University of Liverpool

process, or system lifecycle development is the appropriate context for an engineering education. In Chapter Nine, six of the other standards were identified as being the distinguishing features of a CDIO program. The remaining five are considered supplementary, supporting the adoption of best practice.

For educational programs that do not embrace the entire approach, relevant parts can be applied. The CDIO approach then becomes a collection of tools for program development and teaching support. For example, programs that do not accept the lifecycle context or the key role of design-implement experiences may see benefits in the systematic approach toward program development. The focus on systematic planning and documentation, stakeholder engagement, peer comparison, and modern workspaces may be perceived as new and useful.

Application to graduate programs

The CDIO Initiative began as a program for reform of undergraduate engineering education. There is broad interest in adapting a CDIO approach to Master level programs, especially in European contexts. Increasingly, doctoral programs that aim to develop project management and communications skills as well as research skills are emerging, in particular with the intention of educating "doctors for industry." To answer the question of how the approach can be applied in a three-tiered educational system, one must keep in mind the essential aspects, not the details of implementation. A CDIO program provides an education within a context of professional engineering. This education is characterized by educational goals, set by stakeholders, met by sequences of experiential learning activities, and embedded in an integrated curriculum of mutually supporting disciplines.

The professional role of engineers as context. It is evident that this is a factor that may vary from Bachelor to doctoral degrees. While most Bachelor programs aim primarily to educate engineers, most doctoral programs aim primarily to educate researchers. For Master degree programs, there is a full range, from research-oriented programs to engineering-oriented. In order to be able to accommodate these variations, the context may be generalized from the "role of professional engineers" to the "role of professionals," the latter enabling programs to make a deliberate decision as to whether the professional context is research or engineering. Variations may lead to context definitions such as, "*The X program is strongly research-oriented where students learn how to think, analyze, and solve problems in a research context rather than in the technical production context. The emphasis is more on knowledge production than on "product production*" [8]. Such a modification of the context leads to changes with respect to other CDIO Standards, but many are still applicable, as discussed below.

Educational goals set by stakeholders and met by proper sequence of learning activities. This topic prompts the question of what parts of the CDIO Syllabus are applicable for Master and doctoral programs and which are not. In addition, the question, *What increased level of proficiency is expected on the Master and doctoral levels?* is discussed.

Beginning with the question of scope, we argue that Sections 2 and 3 of the Syllabus list knowledge and skills that are important for researchers as well as for engineers. It is evident that researchers need personal skills, such as problem solving, experimentation, knowledge discovery, and systems thinking. Interpersonal skills are equally important for research work. Current research is typically conducted in international teams requiring an ability to cooperate. To be successful in acquiring research funding not only requires good ideas, but also communication skills. Learning outcomes for communication skills in a research-oriented program may be specialized toward research-related communication tasks, for example, writing journal articles and research proposals. The headings in Section 4—Conceiving, Designing, Implementing, Operating—may be more or less applicable depending on the Master or doctoral program. A program in product development may opt to use all headings in Section 4, while a physics program may opt to use none. Regardless of the selection of appropriate parts of the Syllabus, there are tools for writing program goal statements and course learning outcomes.

The next issue to consider is the difference of skill outcomes at end of the Bachelor, Master and doctoral degrees. What increased level of proficiency in skills is expected for the Master degree level? These differences have not been quantitatively investigated in the Initiative so far, and remain an open issue. Pending additional studies, some indications are likely to emerge during the Bologna Process, such as internationally accepted guidelines for what characterizes Bachelor, Master, and doctoral levels, with respect to some high-level goals, including technical knowledge as well as communication skills [1]. It is also likely that such internationally agreed-upon goals will remain abstract, leaving the details to programs in consultation with their respective stakeholders.

Applying a CDIO approach to a Master degree program that is closely coupled to a Bachelor degree program will also require consideration of the sequence of learning experiences. Specifically, Standard 4 *Introduction to Engineering* and Standard 5 *Design-Implement Experiences* explicitly recommend types of learning experiences in the curriculum. It will be a challenge to work with students who have not had these experiences at the undergraduate level. Such students may be well prepared in terms of technical knowledge, but less capable with regard to personal and interpersonal skills, and product, process, and system building skills. Accommodation of these students in the form of additional learning experiences may be necessary.

Application beyond engineering education

The principles and practices of the CDIO approach can be applied to most programs in higher education. At its most abstract level, the approach asserts the following: the education should be in the context of practice; that there is an identifiable list of knowledge, skills, and attitudes in which students should gain proficiency; that by engaging with stakeholders, the desired level of proficiency can be determined; that curriculum and pedagogy should be constructed in an integrated manner to reasonably ensure meeting the desired learning outcomes; and that learner assessment and program evaluation should be aligned with learning outcomes that in turn should be used to inform faculty and students of progress, and serve as the basis of continuous improvement. What curriculum would not benefit from systematically applying this approach?

At the next level of detail, the Syllabus, which defines the desired learning outcomes for engineering programs, can be easily adapted to virtually any educational program. Section 1 *Technical Knowledge and Reasoning*, can be changed to *Disciplinary Knowledge and Reasoning*. Sections 2 and 3, *Personal and Interpersonal Knowledge and Skills*, are largely common to all university education. A modification to the description of the product lifecycle in Section *4 Conceiving, Designing, Implementing, and Operating Systems in the Enterprise and Societal Context*, to Applying Knowledge to Benefit Society could generalize that section. (See Figure 3.7 in Chapter 3) Similarly, the Standards can be adapted with a modification to Standard 1 *The Context* to "education in the context of practice".

It is also possible to consider similar professions that aim to use products, processes, or systems to explicitly shape outcomes. Architecture, medicine, education, and business management fall into this category. The underlying processes in the practice of architecture are much like engineering, with perhaps more emphasis on aesthetics and visual design. The adaptation of a CDIO approach in this domain would be the most direct. In fact, many architectural educators might observe that Conceiving-Designing-Implementing-Operating moves engineering education closer to architectural education, with its emphasis on experiential learning.

Standard 1 defines product, process, and system lifecycle development and deployment as the context for engineering education. In engineering, there is usually a clear interpretation of the meaning of a product, process, or a system, but these concepts may not exist in other areas. For example, in medical education or teacher education, the context is one of service to, and improvement of, patients or students. The introduction of the concept of services to the Syllabus and the Standards could facilitate adaptation of the approach beyond engineering to these fields. Business education is another area of potential application. In business and management programs, there is a need for extending or modifying the definitions of products, processes, and systems. However, to the extent that professionals in business management define strategies, organizations, products, and services, much of the CDIO approach is applicable.

Adaptation to fields in which professionals do not explicitly use products, processes, and systems to shape outcomes requires consideration at the most abstract level. Application to the social sciences, humanities, arts and sciences would raise questions such as, *What is the context of practice?* and *Who are the appropriate stakeholders?*

SUMMARY

This chapter has reflected on what we see as the future challenges for engineering education, and the ways in which the CDIO Initiative can contribute to meeting these challenges. Issues included scientific breakthroughs and technological developments, internationalization, student mobility and flexibility, the skills and attitudes of beginning engineering students, issues of gender and widening participation, and governmental policies and initiatives. Finally, we discussed how the CDIO approach can be applied to additional engineering areas, graduate education, and programs beyond engineering.

We began this book with the following paragraph: The purpose of engineering education is to provide the learning required by students to become successful engineers—technical expertise, social awareness, and a bias toward innovation. This combined set of knowledge, skills, and attitudes is essential to strengthening productivity, entrepreneurship, and excellence in an environment that is increasingly based on technologically complex and sustainable products, processes, and systems. It is imperative that we improve the quality and nature of undergraduate engineering education.

We believe that the CDIO Initiative has developed an approach to meeting this imperative. It responds to the identified needs of educating students who are "ready to engineer." It has as its goals that learning should be strengthened both in the fundamentals and skills. It has developed a pragmatic and systematic approach to an integrated curriculum and pedagogy with appropriately aligned assessment tools.

We have produced a set of open resources to make this approach available to others, with the understanding that nothing is prescriptive. We offer a set of resources and approaches that can be adapted and implemented in every local program. We hope that these resources will continue to grow as others contribute.

We anticipate the CDIO Initiative will emerge over the next few years as one of many successful experiments in the reform of engineering education worldwide. We continue to reflect on the outcomes of these efforts to improve the education of our students, as we prepare them to build the technologically complex and sustainable products, processes, and systems that are important to our future.

DISCUSSION QUESTIONS

1. In what ways do you expect engineering education to change in the next ten years? In the next 20 years?
2. What developments in science, technology, and business are likely to have the most influence on engineering education in the next 20 years?
3. In what ways do you expect your own program to change in light of the ideas presented in this book?

References

[1] The Bologna Process. Available at http://europa.eu.int/comm/education/policies/educ/bologna/bologna_en.html

[2] Canadian Engineering Accreditation Board, Canadian Council of Professional Engineers. Available at http://www.ccpe.ca

[3] Washington Accord. Available at http://www.washingtonaccord.org

[4] American Board of Engineering and Technology, *Criteria for Accrediting Engineering Programs Effective for Evaluations During the 2000-2001 Accreditation Cycle*, 2000. Available at http://www.abet.org

[5] Accreditation of European Engineering Programmes. Available at http://www.eurace.org

[6] Augusti. G., *Accreditation of Engineering Programmes: A Pan-European Approach.* 33rd SEFI Annual Conference, Ankara, Turkey, September 7-10, 2005.

[7] Irandoust, S. et al., *Att lyfta matematiken–intresse, lärande, kompetens* (Lifting Mathematics–Interest, Learning, Competence), SOU2004:97, 2004. In Swedish.
[8] Malmqvist, J., Edström, K., Gunnarsson, S., Östlund, S., "*Use of CDIO Standards in Swedish National Evaluation of Engineering Educational Programs*", Proceedings of 1st CDIO Conference, Kingston, Canada, 2005.
[9] Franzén, C. J., Private communication, Department of Chemical and Biological Engineering, Chalmers University of Technology, Göteborg, Sweden., 2005.

THE CDIO SYLLABUS

1 TECHNICAL KNOWLEDGE AND REASONING

1.1 KNOWLEDGE OF UNDERLYING SCIENCES
1.1.1 (Defined by program)

1.2 CORE ENGINEERING FUNDAMENTAL KNOWLEDGE
1.2.1 (Defined by program)

1.3 ADVANCED ENGINEERING FUNDAMENTAL KNOWLEDGE
1.3.1 (Defined by program)

2 PERSONAL AND PROFESSIONAL SKILLS AND ATTRIBUTES

2.1 ENGINEERING REASONING AND PROBLEM SOLVING
2.1.1 Problem Identification and Formulation
Evaluate data and symptoms
Analyze assumptions and sources of bias
Demonstrate issue prioritization in context of overall goals
Formulate a plan of attack (incorporating model, analytical and numerical solutions, qualitative analysis, experimentation and consideration of uncertainty)
2.1.2 Modeling
Employ assumptions to simplify complex systems and environment
Choose and apply conceptual and qualitative models
Choose and apply quantitative models and simulations
2.1.3 Estimation and Qualitative Analysis
Estimate orders of magnitude, bounds and trends
Apply tests for consistency and errors (limits, units, etc.)
Demonstrate the generalization of analytical solutions
2.1.4 Analysis With Uncertainty
Elicit incomplete and ambiguous information
Apply probabilistic and statistical models of events and sequences

Practice engineering cost-benefit and risk analysis
Discuss decision analysis
Schedule margins and reserves

2.1.5 Solution and Recommendation
Synthesize problem solutions
Analyze essential results of solutions and test data
Analyze and reconcile discrepancies in results
Formulate summary recommendations
Appraise possible improvements in the problem solving process

2.2 EXPERIMENTATION AND KNOWLEDGE DISCOVERY

2.2.1 Hypothesis Formulation
Select critical questions to be examined
Formulate hypotheses to be tested
Discuss controls and control groups

2.2.2 Survey of Print and Electronic Literature
Choose the literature research strategy
Demonstrate information search and identification using library tools (on-line catalogs, databases, search engines)
Demonstrate sorting and classifying the primary information
Question the quality and reliability of information
Identify the essentials and innovations contained in the information
Identify research questions that are unanswered
List citations to references

2.2.3 Experimental Inquiry
Formulate the experimental concept and strategy
Discuss the precautions when humans are used in experiments
Execute experiment construction
Execute test protocols and experimental procedures
Execute experimental measurements
Analyze and report experimental data
Compare experimental data vs. available models

2.2.4 Hypothesis Test, and Defense
Discuss the statistical validity of data
Discuss the limitations of data employed
Prepare conclusions, supported by data, needs and values
Appraise possible improvements in knowledge discovery process

2.3 SYSTEM THINKING

2.3.1 Thinking Holistically
Identify and define a system, its behavior, and its elements

Use trans-disciplinary approaches that ensure the system is understood from all relevant perspectives
Identify the societal, enterprise and technical context of the system
Identify the interactions external to the system, and the behavioral impact of the system

2.3.2 Emergence and Interactions in Systems
Discuss the abstractions necessary to define and model system
Identify the behavioral and functional properties (intended and unintended) which emerge from the system
Identify the important interfaces among elements
Recognize evolutionary adaptation over time

2.3.3 Prioritization and Focus
Locate and classify all factors relevant to the system in the whole
Identify the driving factors from among the whole
Explain resource allocations to resolve the driving issues

2.3.4 Trade-offs, Judgment and Balance in Resolution
Identify tensions and factors to resolve through trade-offs
Choose and employ solutions that balance various factors, resolve tensions and optimize the system as a whole
Describe flexible vs. optimal solutions over the system lifetime
Appraise possible improvements in the system thinking used

2.4 PERSONAL SKILLS AND ATTITUDES

2.4.1 Initiative and Willingness to Take Risks
Identify the needs and opportunities for initiative
Discuss the potential benefits and risks of an action
Explain the methods and timing of project initiation
Demonstrates leadership in new endeavors, with a bias for appropriate action
Practice definitive action, delivery of results and reporting on actions

2.4.2 Perseverance and Flexibility
Demonstrate self-confidence, enthusiasm, and passion
Demonstrate the importance of hard work, intensity and attention to detail
Demonstrate adaptation to change
Demonstrate a willingness and ability to work independently
Demonstrate a willingness to work with others, and to consider and embrace various viewpoints

Demonstrate an acceptance of criticism and positive response
Discuss the balance between personal and professional life

2.4.3 Creative Thinking
Demonstrate conceptualization and abstraction
Demonstrate synthesis and generalization
Execute the process of invention
Discuss the role of creativity in art, science, the humanities and technology

2.4.4 Critical Thinking
Analyze the statement of the problem
Choose logical arguments and solutions
Evaluate supporting evidence
Locate contradictory perspectives, theories and facts
Identify logical fallacies
Test hypotheses and conclusions

2.4.5 Awareness of One's Personal Knowledge, Skills and Attitudes
Describe one's skills, interests, strengths, weaknesses
Discuss the extent of one's abilities, and one's responsibility for self-improvement to overcome important weaknesses
Discuss the importance of both depth and breadth of knowledge

2.4.6 Curiosity and Lifelong Learning
Discuss the motivation for continued self-education
Demonstrate the skills of self-education
Discuss one's own learning style
Discuss developing relationships with mentors

2.4.7 Time and Resource Management
Discuss task prioritization
Explain the importance and/or urgency of tasks
Explain efficient execution of tasks

2.5 PROFESSIONAL SKILLS AND ATTITUDES

2.5.1 Professional Ethics, Integrity, Responsibility & Accountability
Demonstrate one's ethical standards and principles
Demonstrate the courage to act on principle despite adversity
Identify the possibility of conflict between professionally ethical imperatives
Demonstrate an understanding that it is acceptable to make mistakes, but that one must be accountable for them
Practice proper allocation of credit to collaborators
Demonstrate a commitment to service

2.5.2 Professional Behavior
Discuss a professional bearing
Explain professional courtesy
Identify international customs and norms of interpersonal contact
2.5.3 Proactively Planning for One's Career
Discuss a personal vision for one's future
Explain networks with professionals
Identify one's portfolio of professional skills
2.5.4 Staying Current on World of Engineer
Discuss the potential impact of new scientific discoveries
Describe the social and technical impact of new technologies and innovations
Discuss a familiarity with current practice/technology in engineering
Explain the links between engineering theory and practice

3 INTERPERSONAL SKILLS: TEAMWORK AND COMMUNICATION

3.1 TEAMWORK

3.1.1 Forming Effective Teams
Identify the stages of team formation and life cycle
Interpret task and team processes
Identify team roles and responsibilities
Analyze the goals, needs and characteristics (works styles, cultural differences) of individual team members
Analyze the strengths and weakness of the team
Discuss ground rules on norms of team confidentiality, accountability and initiative
3.1.2 Team Operation
Choose goals and agenda
Execute the planning and facilitation of effective meetings
Apply team ground rules
Practice effective communication (active listening, collaboration, providing and obtaining information)
Demonstrate positive and effective feedback
Practice the planning, scheduling and execution of a project
Formulate solutions to problems (creativity and decision making)
Practice conflict negotiation and resolution
3.1.3 Team Growth and Evolution
Discuss strategies for reflection, assessment, and self-assessment
Identify skills for team maintenance and growth

Identify skills for individual growth within the team
Explain strategies for team communication and writing

3.1.4 Leadership
Explain team goals and objectives
Practice team process management
Practice leadership and facilitation styles (directing, coaching, supporting, delegating)
Explain approaches to motivation (incentives, example, recognition, etc)
Practice representing the team to others
Describe mentoring and counseling

3.1.5 Technical Teaming
Describe working in different types of teams
Cross-disciplinary teams (including non-engineer)
Small team vs. large team
Distance, distributed and electronic environments
Demonstrate technical collaboration with team members

3.2 COMMUNICATIONS

3.2.1 Communications Strategy
Analyze the communication situation
Choose a communications strategy

3.2.2 Communications Structure
Construct logical, persuasive arguments
Construct the appropriate structure and relationship amongst ideas
Choose relevant, credible, accurate supporting evidence
Practice conciseness, crispness, precision and clarity of language
Analyze rhetorical factors (e.g. audience bias)
Identify cross-disciplinary cross-cultural communications

3.2.3 Written Communication
Demonstrate writing with coherence and flow
Practice writing with correct spelling, punctuation and grammar
Demonstrate formatting the document
Demonstrate technical writing
Apply various written styles (informal, formal memos, reports, etc)

3.2.4 Electronic/Multimedia Communication
Demonstrate preparing electronic presentations
Identify the norms associated with the use of e-mail, voice mail, and videoconferencing
Apply various electronic styles (charts, web, etc)

3.2.5 Graphical Communication
- Demonstrate sketching and drawing
- Demonstrate construction of tables, graphs and charts
- Interpret formal technical drawings and renderings

3.2.6 Oral Presentation and Inter-Personal Communications
- Practice preparing presentations and supporting media with appropriate language, style, timing and flow
- Use appropriate nonverbal communications (gestures, eye contact, poise)
- Demonstrate answering questions effectively

3.3 COMMUNICATIONS IN FOREIGN LANGUAGES

3.3.1 English
3.3.2 Languages of regional industrial nations
3.3.3 Other languages

4 CONCEIVING, DESIGNING, IMPLEMENTING AND OPERATING SYSTEMS IN THE ENTERPRISE AND SOCIETAL CONTEXT

4.1 EXTERNAL AND SOCIETAL CONTEXT

4.1.1 Roles and Responsibility of Engineers
- Accepts the goals and roles of the engineering profession
- Accepts the responsibilities of engineers to society

4.1.2 The Impact of Engineering on Society
- Explain the impact of engineering on the environment, social, knowledge and economic systems in modern culture

4.1.3 Society's Regulation of Engineering
- Accepts the role of society and its agents to regulate engineering
- Recognize the way in which legal and political systems regulate and influence engineering
- Describe how professional societies license and set standards
- Describe how intellectual property is created, utilized and defended

4.1.4 The Historical and Cultural Context
- Describe the diverse nature and history of human societies as well as their literary, philosophical, and artistic traditions
- Describe the discourse and analysis appropriate to the discussion of language, thought and values

4.1.5 Contemporary Issues and Values
- Describe the important contemporary political, social, legal and environmental issues and values

Define the process by which contemporary values are set, and one's role in these processes
Define the mechanisms for expansion and diffusion of knowledge

4.1.6 Developing a Global Perspective
Describe the internationalization of human activity
Recognize the similarities and differences in the political, social, economic, business and technical norms of various cultures
Recognize international inter-enterprise and inter-governmental agreements and alliances

4.2 ENTERPRISE AND BUSINESS CONTEXT

4.2.1 Appreciating Different Enterprise Cultures
Recognize the differences in process, culture, and metrics of success in various enterprise cultures:
Corporate vs. academic vs. governmental vs. non-profit/NGO
Market vs. policy driven
Large vs. small
Centralized vs. distributed
Research and development vs. operations
Mature vs. growth phase vs. entrepreneurial
Longer vs. faster development cycles
With vs. without the participation of organized labor

4.2.2 Enterprise Strategy, Goals, and Planning
State the mission and scope of the enterprise
Recognize an enterprise's core competence and markets
Recognize the research and technology process
Recognize key alliances and supplier relations
List financial and managerial goals and metrics
Recognize financial planning and control
Describe stake-holder relations (with owners, employees, customers, etc.)

4.2.3 Technical Entrepreneurship
Recognize entrepreneurial opportunities that can be addressed by technology
Recognize technologies that can create new products and systems
Describe entrepreneurial finance and organization

4.2.4 Working Successfully in Organizations
Define the function of management
Describe various roles and responsibilities in an organization
Describe the roles of functional and program organizations

Describe working effectively within hierarchy and organizations
Describe change, dynamics and evolution in organizations

4.3 CONCEIVING AND ENGINEERING SYSTEMS

4.3.1 Setting System Goals and Requirements

Identify market needs and opportunities
Elicit and interpret customer needs
Identify opportunities that derive from new technology or latent needs
Explain factors that set the context of the requirements
Identify enterprise goals, strategies, capabilities and alliances
Locate and classify competitors and benchmarking information
Interpret ethical, social, environmental, legal and regulatory influences
Explain the probability of change in the factors that influence the system, its goals and resources available
Interpret system goals and requirements
Identify the language/format of goals and requirements
Interpret initial target goals (based on needs, opportunities and other influences)
Explain system performance metrics
Interpret requirement completeness and consistency

4.3.2 Defining Function, Concept and Architecture

Identify necessary system functions (and behavioral specifications)
Select system concepts
Identify the appropriate level of technology
Analyze trade-offs among and recombination of concepts
Identify high level architectural form and structure
Discuss the decomposition of form into elements, assignment of function to elements, and definition of interfaces

4.3.3 Modeling of System and Ensuring Goals Can Be Met

Locate appropriate models of technical performance
Discuss the concept of implementation and operations
Discuss life cycle value and costs (design, implementation, operations, opportunity, etc.)
Discuss trade-offs among various goals, function, concept and structure and iteration until convergence

4.3.4 Development Project Management

Describe project control for cost, performance, and schedule

Explain appropriate transition points and reviews
Explain configuration management and documentation
Interpret performance compared to baseline
Define earned value process
Discuss the estimation and allocation of resources
Identify risks and alternatives
Describe possible development process improvements

4.4 DESIGNING

4.4.1 The Design Process

Choose requirements for each element or component derived from system level goals and requirements
Analyze alternatives in design
Select the initial design
Use prototypes and test articles in design development
Execute appropriate optimization in the presence of constraints
Demonstrate iteration until convergence
Synthesize the final design
Demonstrate accommodation of changing requirements

4.4.2 The Design Process Phasing and Approaches

Explain the activities in the phases of system design (e.g. conceptual, preliminary, and detailed design)
Discuss process models appropriate for particular development projects (waterfall, spiral, concurrent, etc.)
Discuss the process for single, platform and derivative products

4.4.3 Utilization of Knowledge in Design

Utilize technical and scientific knowledge
Practice creative and critical thinking, and problem solving
Discuss prior work in the field, standardization and reuse of designs (including reverse engineer and redesign)
Discuss design knowledge capture

4.4.4 Disciplinary Design

Choose appropriate techniques, tools, and processes
Explain design tool calibration and validation
Practice quantitative analysis of alternatives
Practice modeling, simulation and test
Discuss analytical refinement of the design

4.4.5 Multidisciplinary Design

Identify interactions between disciplines
Identify dissimilar conventions and assumptions
Explain differences in the maturity of disciplinary models
Explain multidisciplinary design environments
Explain multidisciplinary design

4.4.6 Multi-Objective Design (DFX)
Demonstrate design for:
Performance, life cycle cost and value
Aesthetics and human factors
Implementation, verification, test and environmental sustainability
Operations
Maintainability, reliability, and safety
Robustness, evolution, product improvement and retirement

4.5 IMPLEMENTING

4.5.1 Designing the Implementation Process
State the goals and metrics for implementation performance, cost and quality
Recognize the implementation system design:

4.5.2 Hardware Manufacturing Process
Describe the manufacturing of parts
Describe the assembly of parts into larger constructs
Define tolerances, variability, key characteristics and statistical process control

4.5.3 Software Implementing Process
Explain the break down of high-level components into module designs (including algorithms and data structures)
Discuss algorithms (data structures, control flow, data flow)
Describe the programming language
Execute the low-level design (coding)
Describe the system build

4.5.4 Hardware Software Integration
Describe the integration of software in electronic hardware (size of processor, communications, etc)
Describe the integration of software integration with sensor, actuators and mechanical hardware
Describe hardware/software function and safety

4.5.5 Test, Verification, Validation, and Certification
Discuss test and analysis procedures (hardware vs. software, acceptance vs. qualification)
Discuss the verification of performance to system requirements
Discuss the validation of performance to customer needs
Explain the certification to standards

4.5.6 Implementation Management
Describe the organization and structure for implementation

Discuss sourcing, partnering, and supply chains
Recognize control of implementation cost, performance and schedule
Describe quality and safety assurance
Describe possible implementation process improvements

4.6 OPERATING

4.6.1 Designing and Optimizing Operations
Interpret the goals and metrics for operational performance, cost, and value
Explain operations process architecture and development
Explain operations (and mission) analysis and modeling

4.6.2 Training and Operations
Describe training for professional operations:
Simulation
Instruction and programs
Procedures
Recognize education for consumer operation
Describe operations processes
Recognize operations process interactions

4.6.3 Supporting the System Lifecycle
Explain maintenance and logistics
Describe lifecycle performance and reliability
Describe lifecycle value and costs
Explain feedback to facilitate system improvement

4.6.4 System Improvement and Evolution
Define pre-planned product improvement
Recognize improvements based on needs observed in operation
Recognize evolutionary system upgrades
Recognize contingency improvements/solutions resulting from operational necessity

4.6.5 Disposal and Life-End Issues
Define the end of life issues
List disposal options
Define residual value at life-end
List environmental considerations for disposal

4.6.6 Operations Management
Describe the organization and structure for operations
Recognize partnerships and alliances
Recognize control of operations cost, performance and scheduling
Describe quality and safety assurance
Define life cycle management
Recognize possible operations process improvements

THE CDIO STANDARDS

BACKGROUND

A major international project to reform undergraduate engineering education was launched in October 2000. This project, called *The CDIO Initiative*, has expanded to include engineering programs worldwide. The vision of the project is to provide students with an education that stresses engineering fundamentals set in the context of Conceiving–Designing–Implementing–Operating (CDIO) real-world systems and products. The *CDIO Initiative* has three overall goals - to educate students who are able to:

- master a deep working knowledge of technical fundamentals
- lead in the creation and operation of new products and systems
- understand the importance and strategic impact of research and technological development on society

The *CDIO Initiative* creates resources that can be adapted and implemented by individual programs to meet these goals. These resources support a curriculum organized around mutually supporting disciplines, interwoven with learning experiences related to personal and interpersonal skills, and product, process, and system building skills. Students receive an education rich in design-implement experiences and active and experiential learning, set in both the classroom and modern learning workspaces. One of these resources, the CDIO Standards, is provided in this document. For more information about the *CDIO Initiative*, visit http://www.cdio.org

THE CDIO STANDARDS

In January 2004, the *CDIO Initiative* adopted 12 standards that describe CDIO programs. These guiding principles were developed in response to program leaders, alumni, and industrial partners who wanted to know how they would recognize CDIO programs and their graduates. As a result, these CDIO Standards define the distinguishing features of a CDIO program, serve as guidelines for educational program reform and evaluation, create benchmarks and goals with worldwide application, and provide a framework for continuous improvement.

The 12 CDIO Standards address program philosophy (Standard 1), curriculum development (Standards 2, 3 and 4), design-implement experiences and workspaces (Standards 5 and 6), new methods of teaching and learning (Standards 7 and 8), faculty development (Standards 9 and 10), and assessment and evaluation (Standards 11 and 12). Of these 12 standards, seven are considered *essential* because they distinguish CDIO programs from other educational reform initiatives. (An asterisk [*] indicates these essential standards.) The five *supplementary* standards significantly enrich a CDIO program and reflect best practice in engineering education.

For each standard, the *description* explains the meaning of the standard, the *rationale* highlights reasons for setting the standard, and *evidence* gives examples of documentation and events that demonstrate compliance with the standard.

STANDARD 1 – THE CONTEXT*

Adoption of the principle that product, process, and system lifecycle development and deployment – Conceiving, Designing, Implementing and Operating – are the context for engineering education

Description: A CDIO program is based on the principle that product, process, and system lifecycle development and deployment are the appropriate context for engineering education. *Conceiving–Designing–Implementing– Operating* is a model of the entire product, process, and system lifecycle. The *Conceive* stage includes defining customer needs; considering technology, enterprise strategy, and regulations; and, developing conceptual, technical, and business plans. The second stage, *Design*, focuses on creating the design, *i.e.*, the plans, drawings, and algorithms that describe what will be implemented. The *Implement* stage refers to the transformation of the design into the product, process, or system, including manufacturing, coding, testing and validation. The final stage, *Operate*, uses the implemented product or process to deliver the intended value, including maintaining, evolving and retiring the system.

The product, process, and system lifecycle is considered the *context* for engineering education in that it is the cultural framework, or environment, in which technical knowledge and other skills are taught, practiced and learned. The principle is *adopted* by a program when there is explicit agreement of faculty to transition to a CDIO program, and support from program leaders to sustain reform initiatives.

Rationale: Beginning engineers should be able to *Conceive–Design–Implement–Operate* complex value-added engineering products, processes, and systems in modern team-based environments. They should be able to participate in engineering processes, contribute to the development of engineering products, and do so while working in engineering organizations. This is the essence of the engineering profession.

Evidence:

- a mission statement, or other documentation approved by appropriate responsible bodies, that describes the program as being a CDIO program
- faculty and students who can explain the principle that the product, process, and system lifecycle is the context of engineering education

> STANDARD 2 – **LEARNING OUTCOMES***
>
> **Specific, detailed learning outcomes for personal and interpersonal skills, and product, process, and system building skills, as well as disciplinary knowledge, consistent with program goals and validated by program stakeholders**

Description: The knowledge, skills, and attitudes intended as a result of engineering education, *i.e.*, the *learning outcomes*, are codified in the *CDIO Syllabus*. These learning outcomes detail what students should know and be able to do at the conclusion of their engineering programs. In addition to learning outcomes for technical disciplinary knowledge (Section 1), the *CDIO Syllabus* specifies learning outcomes as personal and interpersonal skills, and product, process, and system building. *Personal* learning outcomes (Section 2) focus on individual students' cognitive and affective development, for example, engineering reasoning and problem solving, experimentation and knowledge discovery, system thinking, creative thinking, critical thinking, and professional ethics. *Interpersonal* learning outcomes (Section 3) focus on individual and group interactions, such as, teamwork, leadership, and communication. *Product, process, and system building* skills (Section 4) focus on conceiving, designing, implementing, and operating systems in enterprise, business, and societal contexts.

Learning outcomes are reviewed and validated by key *stakeholders*, groups who share an interest in the graduates of engineering programs, for consistency with *program goals* and relevance to engineering practice. In addition, stakeholders help to determine the expected level of proficiency, or standard of achievement, for each learning outcome.

Rationale: Setting specific learning outcomes helps to ensure that students acquire the appropriate foundation for their future. Professional engineering organizations and industry representatives have identified key attributes of beginning engineers both in technical and professional areas. Moreover, many evaluation and accreditation bodies expect engineering programs to identify program outcomes in terms of their graduates' knowledge, skills, and attitudes.

Evidence:

- learning outcomes that include knowledge, skills, and attitudes of graduating engineers
- learning outcomes validated for content and proficiency level by key stakeholders (for example, faculty, students, alumni, and industry representatives)

STANDARD 3 – INTEGRATED CURRICULUM*

A curriculum designed with mutually supporting disciplinary courses, with an explicit plan to integrate personal and interpersonal skills, and product, process, and system building skills

Description: An integrated curriculum includes learning experiences that lead to the acquisition of *personal and interpersonal skills, and product, process, and system building skills* (Standard 2), interwoven with the learning of disciplinary knowledge. Disciplinary courses are *mutually supporting* when they make explicit connections among related and supporting content and learning outcomes. An *explicit plan* identifies ways in which the integration of skills and multidisciplinary connections are to be made, for example, by mapping the specified learning outcomes to courses and co-curricular activities that make up the curriculum.

Rationale: The teaching of personal and interpersonal skills, and product, process, and system building skills should not be considered an addition to an already full curriculum, but an integral part of it. To reach the intended learning outcomes in disciplinary knowledge and skills, the curriculum and learning experiences have to make dual use of available time. Faculty play an active role in designing the integrated curriculum by suggesting appropriate disciplinary linkages, as well as opportunities to address specific skills in their respective teaching areas.

Evidence:

- a documented plan that integrates personal and interpersonal skills and product, process, and system building skills with technical disciplinary skowledge, and that exploits appropriate disciplinary linkages
- inclusion of the specified skills in courses and co-curricular activities
- faculty and student recognition of these skills in the curriculum

STANDARD 4 – INTRODUCTION TO ENGINEERING

An introductory course that provides the framework for engineering practice in product, process, and system building, and introduces essential personal and interpersonal skills

Description: The *introductory* course, usually one of the first required courses in a program, provides a framework for the practice of engineering. This *framework* is a broad outline of the tasks and responsibilities of an engineer, and the use of disciplinary knowledge in executing those tasks. Students engage in the *practice of engineering* through problem solving and simple design exercises, individually and in teams. The course also includes personal and interpersonal skills knowledge, skills, and attitudes that are *essential* at the start of a program to prepare students for more advanced product, process, and system building experiences. For example, students can participate in small team exercises to prepare them for larger development teams.

Rationale: Introductory courses aim to stimulate students' interest in, and strengthen their motivation for, the field of engineering by focusing on the application of relevant core engineering disciplines. Students usually elect engineering programs because they want to build things, and introductory courses can capitalize on this interest. In addition, introductory courses provide an early start to the development of the essential skills described in the *CDIO Syllabus.*

Evidence:

- learning experiences that introduce personal and interpersonal skills, and product, process, and system building skills
- student acquisition of the skills described in Standard 2
- high levels of student interest in their chosen field of study, demonstrated, for example, in surveys or choices of subsequent elective courses

STANDARD 5 – **DESIGN-IMPLEMENT EXPERIENCES***

A curriculum that includes two or more design-implement experiences, including one at a basic level and one at an advanced level

Description: The term *design-implement experience* denotes a range of engineering activities central to the process of developing new products and systems. Included are all of the activities described in Standard One at the *Design* and *Implement* stages, plus appropriate aspects of conceptual design from the *Conceive* stage. Students develop product, process, and system building skills, as well as the ability to apply engineering science, in design-implement experiences integrated into the curriculum. Design-implement experiences are considered *basic* or *advanced* in terms of their scope, complexity, and sequence in the program. For example, simpler products and systems are included earlier in the program, while more complex design-implement experiences appear in later courses designed to help students integrate knowledge and skills acquired in preceding courses and learning activities. Opportunities to conceive, design, implement, and operate products, processes, and systems may also be included in required co-curricular activities, for example, undergraduate research projects and internships.

Rationale: Design-implement experiences are structured and sequenced to promote early success in engineering practice. Iteration of design-implement experiences and increasing levels of design complexity reinforce students' understanding of the product, process, and system development process. Design-implement experiences also provide a solid foundation upon which to build deeper conceptual understanding of disciplinary skills. The emphasis on building products and implementing processes in real-world contexts gives students opportunities to make connections between the technical content they are learning and their professional and career interests.

Evidence:

- two or more required design-implement experiences in the curriculum (for example, as part of an introductory course and an advanced course)
- required co-curricular opportunities for design-implement experiences (such as, research labs or internships)
- concrete learning experiences that provide the foundation for subsequent learning of disciplinary skills

STANDARD 6 – ENGINEERING WORKSPACES

Engineering workspaces and laboratories that support and encourage hands-on learning of product, process, and system building, disciplinary knowledge, and social learning

Description: The physical learning environment includes traditional learning spaces, for example, classrooms, lecture halls, and seminar rooms, as well as engineering *workspaces* and *laboratories*. Workspaces and laboratories support the learning of *product, process, and system building skills* concurrently with *disciplinary knowledge*. They emphasize *hands-on learning* in which students are directly engaged in their own learning, and provide opportunities for *social learning*, that is, settings where students can learn from each other and interact with several groups. The creation of new workspaces, or remodeling of existing laboratories, will vary with the size of the program and resources of the institution.

Rationale: Workspaces and other learning environments that support hands-on learning are fundamental resources for learning to design, implement, and operate products, processes, and systems. Students who have access to modern engineering tools, software, and laboratories have opportunities to develop the knowledge, skills, and attitudes that support product, process, and system building competencies. These competencies are best developed in workspaces that are student-centered, user-friendly, accessible, and interactive.

Evidence:

- adequate spaces equipped with modern engineering tools
- workspaces that are student-centered, user-friendly, accessible, and interactive
- high levels of faculty, staff, and student satisfaction with the workspaces

STANDARD 7 – INTEGRATED LEARNING EXPERIENCES*

Integrated learning experiences that lead to the acquisition of disciplinary knowledge, as well as personal and interpersonal skills, and product, process, and system building skills

Description: Integrated learning experiences are pedagogical approaches that foster the learning of disciplinary knowledge simultaneously with personal and interpersonal skills, and product, process, and system building skills. They incorporate professional engineering issues in contexts where they coexist with disciplinary issues. For example, students might consider the analysis of a product, the design of the product, and the social responsibility of the designer of the product, all in one exercise. Industrial partners, alumni, and other key stakeholders are often helpful in providing examples of such exercises.

Rationale: The curriculum design and learning outcomes, prescribed in Standards 2 and 3 respectively, can be realized only if there are corresponding pedagogical approaches that make dual use of student learning time. Furthermore, it is important that students recognize engineering faculty as role models of professional engineers, instructing them in disciplinary knowledge, personal and interpersonal skills, and product, process, and system building skills. With integrated learning experiences, faculty can be more effective in helping students apply disciplinary knowledge to engineering practice and better prepare them to meet the demands of the engineering profession.

Evidence:

- integration of personal and interpersonal skills, and product, process, and system building skills, with disciplinary knowledge in learning activities and experiences
- direct involvement of engineering faculty in implementing integrated learning experiences
- involvement of industrial partners and other stakeholders in the design of learning experiences

STANDARD 8 – **ACTIVE LEARNING**

Teaching and learning based on active experiential learning methods

Description: Active learning methods engage students directly in thinking and problem solving activities. There is less emphasis on passive transmission of information, and more on engaging students in manipulating, applying, analyzing, and evaluating ideas. Active learning in lecture-based courses can include such methods as partner and small-group discussions, demonstrations, debates, concept questions, and feedback from students about what they are learning. Active learning is considered *experiential* when students take on roles that simulate professional engineering practice, for example, design-implement projects, simulations, and case studies.

Rationale: By engaging students in thinking about concepts, particularly new ideas, and requiring some kind of overt response, students not only learn more, they recognize for themselves what and how they learn. This process of metacognition helps to increase students' motivation to achieve program learning outcomes and form habits of lifelong learning. With active learning methods, instructors can help students make connections among key concepts and facilitate the application of this knowledge to new settings.

Evidence:

- successful implementation of active learning methods, documented, for example, by observation or self-report

- a majority of instructors using active learning methods
- high levels of student achievement of all learning outcomes
- high levels of student satisfaction with learning methods

STANDARD 9 – ENHANCEMENT OF FACULTY SKILLS COMPETENCE*

Actions that enhance faculty competence in personal and interpersonal skills, and product, process, and system building skills

Description: CDIO programs provide support for faculty to improve their own competence in the *personal and interpersonal skills, and product, process, and system building skills* described in Standard 2. They develop these skills best in contexts of professional engineering practice. The nature and scope of faculty development vary with the resources and intentions of different programs and institutions. Examples of *actions that enhance faculty competence* include: professional leave to work in industry, partnerships with industry colleagues in research and education projects, inclusion of engineering practice as a criterion for hiring and promotion, and appropriate professional development experiences at the university.

Rationale: If faculty are expected to teach a curriculum of personal and interpersonal skills, and product, process, and system building skills integrated with disciplinary knowledge, as described in Standards 3, 4, 5, and 7, they need to be competent in those skills themselves. Many engineering professors tend to be experts in the research and knowledge base of their respective disciplines, with only limited experience in the practice of engineering in business and industrial settings. Moreover, the rapid pace of technological innovation requires continuous updating of engineering skills. Faculty need to enhance their engineering knowledge and skills so that they can provide relevant examples to students and also serve as role models of contemporary engineers.

Evidence:

- majority of faculty with competence in personal and interpersonal skills, and product, process, and system building skills, demonstrated, for example, by observation and self-report
- high number of faculty with experience in engineering practice
- university's acceptance of professional development in these skills in its faculty evaluation and hiring policies and practices
- commitment of resources for faculty development in these skills

STANDARD 10 – ENHANCEMENT OF FACULTY TEACHING COMPETENCE

Actions that enhance faculty competence in providing integrated learning experiences, in using active experiential learning methods, and in assessing student learning

Description: A CDIO program provides support for faculty to improve their competence in *integrated learning experiences* (Standard 7), active and experiential learning (Standard 8), and assessing student learning (Standard 11). The nature and scope of faculty development practices will vary with programs and institutions. Examples of *actions that enhance faculty competence* include: support for faculty participation in university and external faculty development programs, forums for sharing ideas and best practices, and emphasis in performance reviews and hiring on effective teaching methods.

Rationale: If faculty members are expected to teach and assess in new ways, as described in Standards 7, 8, and 11, they need opportunities to develop and improve these competencies. Many universities have faculty development programs and services that might be eager to collaborate with faculty in CDIO programs. In addition, if CDIO programs want to emphasize the importance of teaching, learning, and assessment, they must commit adequate resources for faculty development in these areas.

Evidence:

- majority of faculty with competence in teaching, learning, and assessment methods, demonstrated, for example, by observation and self-report
- university's acceptance of effective teaching in its faculty evaluation and hiring policies and practices
- commitment of resources for faculty development in these skills

STANDARD 11 – **LEARNING ASSESSMENT***

Assessment of student learning in personal and interpersonal skills, and product, process, and system building skills, as well as in disciplinary knowledge

Description: Assessment of student learning is the measure of the extent to which each student achieves specified learning outcomes. Instructors usually conduct this assessment within their respective courses. Effective learning assessment uses a variety of methods matched appropriately to learning outcomes that address *disciplinary knowledge,* as well as *personal and interpersonal skills, and product, process, and system building skills*, as described in Standard 2. These methods may include written and oral tests, observations of student performance, rating scales, student reflections, journals, portfolios, and peer and self-assessment.

Rationale: If we value personal and interpersonal skills, and product, process, and system building skills, and incorporate them into curriculum and learning experiences, then we must have effective assessment processes for measuring them. Different categories of learning outcomes require different assessment methods. For example, learning outcomes related to *disciplinary knowledge* may be assessed with oral and written tests, while those related to design-implement skills may be better measured with recorded observations. Using a variety of assessment methods accommodates a broader range of learning styles, and increases the reliability and validity of the assessment data. As a result, determinations of students' achievement of the intended learning outcomes can be made with greater confidence.

Evidence:

- assessment methods matched appropriately to all learning outcomes
- successful implementation of assessment methods
- high number of instructors using appropriate assessment methods
- determination of student achievement based on reliable and valid data

STANDARD 12 – **PROGRAM EVALUATION**

A system that evaluates programs against these twelve standards, and provides feedback to students, faculty, and other stakeholders for the purposes of continuous improvement

Description: Program evaluation is a judgment of the overall value of a program based on evidence of a program's progress toward attaining its goals. A CDIO program should be evaluated relative to *these 12 CDIO Standards*. Evidence of overall program value can be collected with course evaluations, instructor reflections, entry and exit interviews, reports of external reviewers, and follow-up studies with graduates and employers. The evidence can be regularly reported back to instructors, students, program administrators, alumni, and other key stakeholders. This *feedback* forms the basis of decisions about the program and its plans for *continuous improvement.*

Rationale: A key function of program evaluation is to determine the program's effectiveness and efficiency in reaching its intended goals. Evidence collected during the program evaluation process also serves as the basis of continuous program improvement. For example, if in an exit interview, a majority of students reported that they were not able to meet some specific learning outcome, a plan could be initiated to identify root causes and implement changes. Moreover, many external evaluators and accreditation bodies require regular and consistent program evaluation.

Evidence:

- a variety of program evaluation methods used to gather data from students, instructors, program leaders, alumni, and other key stakeholders
- a documented continuous improvement process based on results of the program evaluation
- data-driven changes as part of a continuous improvement process

INDEX

Printed in the United States of America.

RECEIVED
OCT 16 2007